Encyclopedia of Alternative and Renewable Energy: Biomass Cultivation and Aquatic Biomass

Volume 08

Encyclopedia of Alternative and Renewable Energy: Biomass Cultivation and Aquatic Biomass

Volume 08

Edited by **Brad Hill and**
David McCartney

New York

Published by Callisto Reference,
106 Park Avenue, Suite 200,
New York, NY 10016, USA
www.callistoreference.com

**Encyclopedia of Alternative and Renewable Energy: Biomass
Cultivation and Aquatic Biomass
Volume 08**
Edited by Brad Hill and David McCartney

International Standard Book Number: 978-1-63239-182-7 (Hardback)

Printed in the United States of America.

Contents

Preface

I am honored to present to you this unique book which encompasses the most up-to-date data in the field. I was extremely pleased to get this opportunity of editing the work of experts from across the globe. I have also written papers in this field and researched the various aspects revolving around the progress of the discipline. I have tried to unify my knowledge along with that of stalwarts from every corner of the world, to produce a text which not only benefits the readers but also facilitates the growth of the field.

Research in the field of biomass and related disciplines has increased with time. Awareness regarding pros and cons of biomass, sustainable use of resources and biomass sources has brought in the concept of biorefineries. This is evident from growth in biomass research and increasing attention towards biofuels. This book covers various disciplines of biomass further expanding on cultivation methods for energy, the feedstock crops and microbial biomass production.

Finally, I would like to thank all the contributing authors for their valuable time and contributions. This book would not have been possible without their efforts. I would also like to thank my friends and family for their constant support.

Editor

Biomass Cultivation

Crop Rotation Biomass and Effects on Sugarcane Yield in Brazil

Edmilson José Ambrosano, Heitor Cantarella, Gláucia Maria Bovi Ambrosano, Eliana Aparecida Schammas, Fábio Luis Ferreira Dias, Fabrício Rossi, Paulo Cesar Ocheuze Trivelin, Takashi Muraoka, Raquel Castellucci Caruso Sachs, Rozario Azcón and Juliana Rolim Salomé Teramoto

Additional information is available at the end of the chapter

1. Introduction

Healthy soils are vital to a sustainable environment. They store carbon, produce food and timber, filter water and support wildlife and the urban and rural landscapes. They also preserve records of ecological and cultural past. However, there are increasing signs that the condition of soils has been neglected and that soil loss and damage may not be recoverable [1]. Soil is a vital and largely non-renewable resource increasingly under pressure. The importance of soil protection is recognized internationally.

In order to perform its many functions, it is necessary to maintain soil condition. However, there is evidence that soil may be increasingly threatened by a range of human activities, which may degrade it. The final phase of the degradation process is land desertification when soil loses its capacity to carry out its functions. Among the threats to soil are erosion, a decline in organic matter, local and diffuse contamination, sealing, compaction, a decline in bio-diversity and salinisation.

The authors in [2] made the interesting observation when study of nine great groups of New Zealand soils, that although many soil quality indicators will be different between soil types differing in clay and organic mater contents, land use had an overriding effect on soil quality: agricultural systems could be clearly differentiated from managed and natural forests, and grass land and arable land were also clearly separated.

Regular additions of organic matter improve soil structure, enhance water and nutrient holding capacity, protect soil from erosion and compaction, and support a healthy community of soil organisms. Practices that increase organic matter include: leaving crop residues in the field, choosing crop rotations that include high residue plants, using optimal nutrient and water management practices to grow healthy plants with large amounts of roots and residue, applying manure or compost, using low or no tillage systems, using sod-based rotations, growing perennial forage crops, mulching, and growing cover crops as green manure [3].

Addition to being a frequent addition of organic matter should be diversified. Diversity cropping systems are beneficial for several reasons. Each plant contributes a unique root structure and type of residue to the soil. A diversity of soil organisms can help control pest populations, and a diversity of cultural practices can reduce weed and disease pressures. Diversity across the landscape can be increased by using buffer strips, small fields, or contour strip cropping. Diversity over time can be increased by using long crop rotations. Changing vegetation across the landscape or over time not only increases plant diversity, but also the types of insects, microorganisms, and wildlife that live on your farm [3]. In addition a dedicated management approach is needed to maintain or increase the soil organic matter content.

The incorporation of plant materials to soils, with the objective of maintaining or improving fertility for the subsequent crop is known as green manuring. The inclusion of a legume fallow within a sugarcane cropping cycle is practiced to reduce populations of detrimental soil organisms [5, 6], provide nitrogen (N) through biological fixation [7,8] and for weed suppression [9, 10].

Interest in the use of green manure's biomass has revived because of their role in improving soil quality and their beneficial N and non-N rotation effects [11]. Because of its nitrogen fixation potential, legumes represent an alternative for supplying nutrients, substituting or complementing mineral fertilization in cropping systems involving green manuring. This practice causes changes in soil physical, chemical and biological characteristics, bringing benefits to the subsequent crop both in small-scale cropping systems and in larger commercial areas such as those grown with sugarcane [12].

The area cropped with sugarcane (*Saccharum spp.*) in Brazil shows rapid expansion, with most of the increase for ethanol production. The area cultivated with sugarcane is now 9.6 Mha, with an increase of 5 Mha from 2000 and over 8.6 Mha of fresh sugarcane harvested per year [13]. Sugarcane crops in Brazil are replanted every five to ten years. In southeastern Brazil, the interval between the last sugarcane harvest and the new plantings occurs during the spring-summer season, under high temperature and heavy rainfall (almost 1,000 mm in six months) [14].

Green manure fertilization of the soil with legumes has been recommended before a sugarcane field is replanted. This practice does not imply on losing the cropping season, does not interfere with sugarcane germination, and provides increases in sugarcane and sugar yield, at least during two consecutive cuts [12]. Additionally, it protects the soil against erosion, prevents weed spreading and reduces nematode populations [15, 16]. Legumes usually accumulate large quantities of N and K, the nutrients which are taken up in the highest amounts by the sugarcane plants.

Residue incorporation studies of legumes using ^{15}N label indicate that 10 to 34% of the legume N can be recovered in the subsequent rye or wheat crop, 42% in rice, 24% recovery from Velvet bean by corn crop, around 15% of N recovery from sunn hemp by corn plants in no-till system, 30% by maize [51],and 5% of N recovery from sunn hemp by sugarcane [12], and ranged from 19 to 21% when the recovery was observed from sunn hemp by two sugarcane harvest [17].

The authors suggested that legume residue decomposition provided long-term supply of N for the subsequent crops, by not supplying the nutrient as an immediate source.

There are many benefits to the sugarcane crop of leguminous plants rown in rotation in sugarcane renovation areas; these include the recycling of nutrients taken up from deep soil layers by the rotational crop, which may prevent or decrease leaching losses, and the addition of N from biological fixation. Leguminous plants can accumulate over 5 t ha^{-1} of dry mass during a short period of time during the summer and take up large amounts of N and K. Most of this N comes from the association of legumes with rhizobia. In this way crop rotation with legumes can replace partially or totally N mineral fertilization of sugarcane, at least for the first ratoon [18, 12].

Another important microbial association is that of mycorrhizal fungi and plant roots. These fungi are present in over 80% of plant species [19]. In contrast with the large diversity of plants, which includes sugarcane, that have their roots colonized by mycorrhizas, only 150 fungi species are responsible for that colonization [19]. A crop whose roots are colonized by mycorrhizal fungi can raise the soil mycorrhizal potential which can benefit plants which are responsive to this fungi association and that are cultivated in sequence. This could be particularly useful for the nutritional management of crops in low nutrient, low input–output systems of production [20].

This study evaluated biomass production, N content, characterizing the biomass and the natural colonization of arbuscular mycorrhizal fungi (AMF) of leguminous green manure and sunflower (*Helianthus annuus* L.) in rotation with sugarcane and their effect in the soil. Their effect on stalk and sugar yield and nematode that occur on sugarcane cv. IAC 87-3396 grown subsequently was also studied. The economic balance considered the costs of production and revenues of the rotational crops as well as three or five harvests of sugarcane was too evaluated.

2. Development

Three long term assays were developed during December 4, 1999 to October 10, 2005 to complete this study as following:

The experiments were carried out in Piracicaba, state of São Paulo, Brazil (22°42′ S, 47°38′ W, and 560 m a.s.l.). The soil, classified as an Arenic Hapludult, and a Typic Paleudult, was chemically characterized at different depths after cutting the green manure crop, before the sugarcane first planting. The soil was acidic and had low amounts of nutrients (Table 1), typical of many sugarcane growing areas.

Total monthly rainfall and local temperature were measured at the meteorological station near the experimental site (Figure 1).

Soil characteristics*	Arenic Hapludult		Typic Paleudult	
	0.0-0.2	0.2-0.4	0.0-0.2	0.2-0.4
	Soil depth, m			
pH (0.01mol l^{-1})	4.1	4.0	5.5	5.5
O.M. (g dm^{-3})	26	22	20	19
P (mg dm^{-3})	3	14	13	10
K (mmolc dm^{-3})	0.7	0.5	0.6	0.4
Ca (mmolc dm^{-3})	7	6	33	29
Mg (mmolc dm^{-3})	6	5	21	19
H + Al (mmolc dm^{-3})	50	68	23	25
CEC (mmolc dm^{-3})	64	80	79	74
V %	22	14	68	64

*Adapted from [12 and 16]

Table 1. Soil chemical characteristics before sugarcane planting, in plots without green manure, at depths of 0.0-0.2 and 0.2-0.4 m.

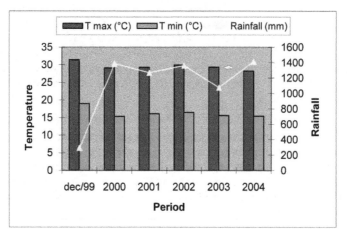

Figure 1. Climatological data of maximum and minimum annual average temperature, and annual average rainfall from December 1999 to December 2004 (experiment 1) adapted from [14].

2.1. Effect of seven rotational crops in sugarcane yield

The experiment 1 consisted to evaluate seven rotational crops plus control (fallow) grown before sugarcane was planted.

The soil is as a Typic Paleudult and was chemically characterized at different depths with samples taken after the green manures were cut but before sugarcane was planted (Table 1).

The experimental design was a randomized block with eight treatments and five replications.

The rotational crops were peanut (*Arachis hypogaea* L.) cv. IAC-Tatu, peanut cv. IAC-Caiapo, sunn hemp cv. IAC-1 (*Crotalaria juncea* L.), velvet bean (*Mucuna aterrima* Piper and Tracy), soybean (*Glycine max* L. Merrill) cv. IAC-17, sunflower (*Helianthus annuus* L.) cv. IAC-Uruguai, and mung bean (*Vigna radiata* L. Wilczek).

The green manures were sowed always in December on 7 m × 10 m size plots, with rows 0.50 m apart. The experimental area was weeded 30 d after sowing, and the weed residues were left on the soil surface.

During seed filling, the plants used as green manure were manually cut and spread on the soil covering the entire plot surface in pieces less than 0.25 m and left there for six months. Peanut, soybean, sunflower and mung bean were harvested after physiological maturation for the grain yield, and the remaining plant parts were cut and spread on the soil. Biomass production of the rotational crops was evaluated in 1 m^2 of the plot area.

At the harvest stage, the roots of each rotational crop were sampled in order to evaluate the natural colonization level of arbuscular mycorrhizal fungi (AMF). The colonization percentage was estimated using the root coloration technique according to [21]. The percentage of colonization by AMF was estimated by counting the roots' stained portions using a reticular plate under a microscope following the procedures described by [22].

To evaluate sugarcane stalk yield 2-m sections of each of the three central rows were cut and weighed.

Ten successive stems were separated from each plot for the technological evaluation of the Brix, pol, and total recovered sugar [23]. Sugar yield, expressed in terms of tons of pol per hectare (TPH), was estimated with the stem yield and technological analysis data.

The economic balance considered the costs of production and revenues of the rotational crops as well as three harvests of sugarcane. The basic costs of production of sugarcane (including land preparation, seed stalk, fertilizer, herbicides feedstock and application, and harvesting) were the average of the 2004, 2005, and 2006 prices, based on an average stalk yield of 70 t ha^{-1}. For the control treatment, which did not include the crop rotations, the cost of production of sugarcane was estimated as U$ 3,111 ha^{-1}. The costs of production of the green manures crotalaria and velvet beans, U$ 100 ha^{-1}, include seeds, planting, and cutting. For the grain crops, the costs of grain harvesting and of chemicals needed for phytosanitary control were added: sunflower (U$ 422 ha^{-1}), peanut cv. IAC-Tatu (U$ 1,289 ha^{-1}), Peanut cv. IAC-Caiapó (U$ 1,480 ha^{-1}), mung bean (U$ 2,007 ha^{-1}) and soybean (U$ 513 ha^{-1}). The sales prices of grain and cane stalks for the period between 2004 and 2006 (according to a database of the Institute of Agricultural Economics of the São Paulo State Secretary of Agriculture) were: sugarcane stalks, U$ 17.56 t^{-1}; sunflower, U$ 178 t^{-1}; peanut cv. IAC-Tatu, U$ 260 t^{-1}; peanut cv. IAC-Caiapó, U$ 260 t^{-1}; soybean, U$ 197 t^{-1}; and mung bean, U$ 2,222 t^{-1}. Mung bean is not sold as a commodity but as a specialty crop; its prices are highly variable, and the market for it is relatively small; therefore, the data on the economical return for mung bean must be taken with care.

2.2. Recovery of nitrogen in sugarcane fertilized with sunn hemp and ammonium sulfate

The utilization of nitrogen by sugarcane (*Saccharum spp.*) fertilized with sunn hemp (SH)(*Crotalaria juncea* L.) and ammonium sulfate (AS) was evaluated using the ^{15}N tracer technique in the experiment 2, that consisted of four treatments with four replications in a randomized block design as fallow: a) control with no N fertilizer or green manure; b) ammonium sulfate (AS) at a rate of 70 kg ha^{-1} N; c) sunn hemp (SH) green manure; d) and sunn hemp plus ammonium sulfate (SH + AS). Microplots consisting of three rows of sugarcane 2-m long were set up in plots c and d with the ^{15}N-labeled sunn hemp.

N was added at the rate of 196 and 70 kg ha^{-1} as ^{15}N labeled sunn hemp green manure (SH) and as ammonium sulfate (AS), respectively. Treatments were: (i) Control; (ii) AS^{15}N; (iii) SH^{15}N + AS; (iv) SH^{15}N; and (v) AS^{15}N + SH. Sugarcane was cultivated for five years and was harvested three times. ^{15}N recovery was evaluated in the two first harvests.

Sunn hemp (*Crotalaria juncea* L, cv IAC-1) was sown at the rate of 25 seeds per meter on the 4 Dec 2000 and emerged in nine days. Microplots, consisting of 6 rows, 2- m long and spaced by 0.5 m within the sunn hemp plots were used for ^{15}N enrichment as described by Ambrosano [24]. After 79 days the sunn hemp was cut, and the fresh material was laid down on the soil surface. Total dry mass of sunn hemp was equivalent to 9.15 Mg ha^{-1}, containing 21.4 g kg^{-1} N, corresponding to 195.8 kg ha^{-1} N with an ^{15}N enrichment of 2.412 atoms % excess.

Microplots with AS-labeled fertilizer (3.01 ± 0.01 atoms % ^{15}N), with two contiguous rows 1-m long, were set up in plots b and also in plots d; therefore, these plots had microplots for both sunn hemp and AS-labeled materials.

Ammonium sulfate was sidedressed to sugarcane 90 days after planting in both main plots and microplots. N rate (70 kg ha^{-1}) is within the range (30 to 90 kg ha^{-1} N) recommended for the plant cane cycle in Brazil [25]. A basal fertilization containing 100 kg ha^{-1} P$_2$O$_5$ as triple superphosphate and 100 kg ha^{-1} K$_2$O as potassium chloride was applied to all treatments to ensure a full sugarcane development. Cane yield was determined outside the microplots by weighing the stalks of three rows of sugarcane, 2-m long.

Stalks yields were measured after 18 months (plant-cane cycle, on 24 Aug 2002), 31 months (1st ratoon crop, on 8 Oct 2003), and 43 months after planting (2nd ratoon crop, on 20 Sep 2004). Samples consisting of ten stalks were used for the determination of apparent sucrose content (Pol) in the cane juice, according to [23]. The expressed cane juice was analyzed for Pol (apparent sucrose) by a saccharimeter. Just before harvesting of the plant cane (24 Aug 2002) and of the first ratoon (8 Oct 2003) whole plants were collected from 1-m row of plants in the center of the microplots. Leaves and stalks were analyzed separately for determination of ^{15}N abundance and N content in a mass spectrometer coupled to an N analyzer, following the methods described in [26].

The fraction and amount of nitrogen in the plant derived from the labeled source (Ndff) and the fraction of N recovery of the labeled source (R%) were calculated based on the isotopic results (atoms %), according to Trivelin [26], Equations 1 to 3:

$$Ndff = (a/b)\,100 \tag{1}$$

$$QNdff = [Ndff\,/100]\,TN \tag{2}$$

$$R\% = [Ndff/\,NF]\,100 \tag{3}$$

where: Ndff (%) is the fraction of nitrogen in the plant derived from the labeled source, a and b are ^{15}N abundance values (atoms % excess) in the plant and in the labeled source (AS or SH), respectively; QNdff (kg ha^{-1}) is the amount of nitrogen in the plant derived from the labeled source, TN (kg ha^{-1}) is total cumulative nitrogen in the sugarcane plant (kg ha^{-1}); R% is the fraction of N recovery of the labeled Sugarcane cultivar IAC- 87-3396 was planted on Mar or April on plots with ten sugarcane rows, 10-m long and spaced at 1.4 m.

The biological nitrogen fixation (BNF) by leguminous plants was determined by natural abundance of ^{15}N technique ($\delta^{15}N$) [27], and sunflower was the non-N fixing specie. The chemical analysis of plants to determine macro and micronutrient contents were performed according to the methods proposed by [28].

2.3. Effect of four rotational crops in sugarcane yield

The experiment 3 consisted to evaluate and characterize the biomass of leguminous residues, the natural arbuscular mycorrhizal (AM) fungus occurrence and the effect of leguminous on the nematodes (*Pratylenchus* spp.) in sugarcane crop. The experiment was carried out in Piracicaba, São Paulo State, Brazil. The soil was classified as Typic Paleudult and the sugarcane (*Saccharum* spp.) cultivar was IAC87-3396. The effects of previous cultivation of legumes were evaluated for five consecutive harvests. The treatments consisted of previous cultivation of legumes: peanut (*Arachis hypogaea* L.) cultivars IAC-Tatu and IAC-Caiapó, sunn hemp IAC 1 (*Crotalaria juncea* L.) and velvet-bean [*Mucuna aterrima* (Piper & Tracy) Holland], and a control treatment. We adopted the randomized block design with five replications.

The chronology of the events on the experimental field in experiment 3 is the same that experiment 1.

3. To evaluate seven rotational crops

The changes in soil properties were relatively small as should be expected with only one rotation. The rotational crops can contribute with organic residues, but, in general, the amounts of organic C added to the soil are usually not enough to cause significant changes in soil organic matter in the short term (Table 2).The letters in the tables represent statistical comparisons. Means followed by at least one equal letter do not differ statistically. Means followed by all the different letters differ significantly. The rotational crops also affected some soil attributes (Table 2). The organic matter content increased in the soil upper layer (0-0.2 m) with the cultivation of peanut cv. IAC-Tatu and velvet bean, and in the 0.2-0.4 m layer, with mung bean, sunflower IAC-Uruguai, and peanut cv. IAC-Tatu. The increase of soil exchangeable magnesium was also observed for peanut cv. IAC-Tatu and velvet bean, although the original Mg content was already high.

Rotational crops	Organic matter			Mg		
	0-0.2 m	0.2-0.4 m	Average	0-0.2 m	0.2-0.4 m	Average
	----------- g kg^{-1} -----------			-------- mmolc dm^{-3} ------		
Control	20 Ab	19 Ab	19	21	19	20 b
Mung bean cv. M146	19 Ab	20 Aa	20	19	18	19 b
Peanut cv. IAC-Caiapó	21 Ab	19 Bb	20	24	15	20 b
Peanut cv. IAC-Tatu	23 Aa	21 Aa	22	29	23	26 a
Soybean cv. IAC-17	19 Ab	17 Bb	18	20	17	18 b
Sunflower cv. IAC-Uruguai	20 Ab	20 Aa	20	20	19	19 b
Sunn hemp IAC 1	19 Ab	18 Ab	18	19	17	18 b
Velvet bean	23 Aa	18 Bb	21	28	18	23 a
Average	21 A	19 B	20	22 A	18 B	20
^{1}C.V.(%)	8.1	8.1		18.4	22.6	

Means followed by the same lower-case letter in the columns and capital letter in the rows are not different (Comparisons among means were made according to Tukey-Kramer test, $p > 0.1$).
^{1}Coefficient of variation. Adapted from [14].

Table 2. Organic matter and exchangeable magnesium in soil sampled after rotational crops.

Sunflower accumulated more above-ground dry matter of total biomass and soybean more grain yield than the other crops (Table 3). Soybean, sunn hemp, velvet bean, and sunflower extracted the greatest amounts of N and P (Table 4). Sunflower also recycled more of K, Ca, Mg, and Zn than the other rotational crops, probably as a consequence of the higher biomass yield (Tables 3 and 4).

Soybean presented the highest N content, and sunflower the lowest. No differences were observed between peanuts and velvet bean and between sunn hemp and mung bean (Table 5). Among the macronutrients, N had the highest and P the lowest accumulation in the rotational crops. On the average Fe was recycled in the highest amounts in the above-ground parts of the rotational crops and Zn in the lowest (Table 4). The same results were observed by [29]) who evaluated pigeon pea (*Cajanus cajan*) and stylo plants (*Stylosanthes guianensis* var. vulgaris cv. Mineirão).

The high AMF infection rate, which helps the uptake of micronutrients (Table 3), may explain the high amounts of Zn returned to the soil when sunflower was grown before sugarcane. There is an increasing utilization of sunflower as a crop rotation with sugarcane in Brazil, due to its use for silage, seed oil production, and to its potential as a feedstock for biodiesel [30].

The amounts of N in the above-ground parts of sunn hemp (Table 4) were relatively low compared to those of [31], who reported the extraction of up to 230 kg ha^{-1} of N, and to those of [12], who found 196 kg ha^{-1} of N. However, the amounts of N returned to the soil are directly related to the nutrient concentration in the plant, which varies with the local potential for biological nitrogen fixation (BNF) and with the growth stage of the crop at the time of cutting, and with the biomass yield, which is affected by the weather, soil, and crop growing conditions.

Rotational crop	Above ground dry matter[2]	Grain yield[2]	Natural infection of AMF
	-------------kg ha^{-1}-----------		%
Control	-	-	-
Mung bean cv. M146	2,225 d	798 d	51 b
Peanut cv. IAC-Caiapó	1,905 d	1,096 c	74 a
Peanut cv. IAC-Tatu	1,783 d	1,349 c	57 b
Soybean cv. IAC-17	3,669 c	2,970 a	56 b
Sunflower cv. IAC-Uruguai	15,229 a	1,805 b	73 a
Sunn hemp IAC 2	6,230 b	-	49 c
Velvet bean	5,049 b	-	65 a
[1]C.V. %	5.1	22.1	15.9

Means followed by the same letter in each column are not different (Comparisons among means were made according to Scott-Knott test, p = 0.05).
[1]Coefficient of variation.
[2] Analysis of variance were made after data transformation to log (×).Adapted from [14].

Table 3. Dry mass and grain yields of the rotational crops and percentage of infection of natural arbuscular mycorrhizal fungus (AMF) in roots of rotational crops.

Rotational crops	N	P	K[2]	Ca[2]	Mg[2]	Fe[2]	Mn[2]	Zn[2]
	------------------------ kg ha^{-1} -----------------					---------------- g ha^{-1} -----------		
Control	-	-	-	-	-	-	-	-
Mung bean cv. M146	27 b	2.4 b	17 d	17 d	12 d	2,073 a	258 a	43 c
Peanut cv. IAC-Caiapo	39 b	2.7 b	35 c	19 d	13 c	3,279 a	222 a	47 c
Peanut cv. IAC-Tatu	34 b	3.8 b	27 c	18 d	11 c	1,679 a	81 b	37 c
Soybean cv. IAC-17	122 a	8.7 a	14 d	46 c	28 b	1,424 a	186 a	56 c
Sunflower IAC-Uruguai	71 a	7.7 a	120 a	171 a	98 a	2,736 a	324 a	259 a
Sunn hemp IAC 2	97 a	5.8 a	33 c	34 c	21 b	1,313 b	178 a	84 b
Velvet bean	109 a	8.9 a	50 b	61 b	17 b	792 a	159 a	90 b
[1]C.V. %	10.9	42.5	13.3	9.7	13.0	9.5	10.1	9.9

Means followed by the same letter in each column are not different (Comparisons among means were made according to Scott-Knott test, p = 0.05).
[1]Coefficient of variation.
[2] Analysis of variance were made after data transformation to log (×).Adapted from [14].

Table 4. Nutrient content of above ground biomass of the rotational crops, excluding the grains.

Perin [32] found substantial amounts of N derived from BNF present in the above ground parts of sunn hemp (57.0%) grown isolated and 61.1% when intercropped with millet (50% seeded with each crop). The sunn hemp+ millet treatment grown before a maize crop resulted in higher grain yield than when sunn hemp alone was the preceding rotation. This effect was not observed when N-fertilizer (90 kg N ha^{-1}) was added. Intercropping legume and cereals is a promising biological strategy to increase and keep N into the production system under tropical conditions [32]. A large proportion of the N present in soybeans usually comes from BNF. Guimarães [33] found that 96% of the N present in above ground parts of soybeans were derived from BNF, values which are in agreement with those obtained by [32] for sunn hemp. However, in the present study, only about 27% of the N present in the soybean residues were from BNF (Table 5), probably because of poor specific population of fixing bacteria for soybeans in the experimental site, which have been grown with sugarcane for long time. No inoculation of soybean with Bradyrhizobium was done. The contribution of BNF for the peanut varieties was significantly different: it reached 70% of the N in the cv. IAC-Caiapó but only 37.7% in the cv. IAC-Tatu (Table 5). Usually the natural population of rhyzobia is high enough to guarantee root colonization for peanuts but probably the bacteria population in the soil of the experimental site was not efficient for peanuts cv. IAC-Tatu.

Rotational crop	C content	N content	C : N	N-BNF
	------------- g kg^{-1} ------------			%
Mung bean cv. M146	426 a	12.5 c	34.1 b	89 a
Peanut cv. IAC-Caiapó	424 a	20.9 b	20.3 b	70 b
Peanut cv. IAC-Tatu	440 a	19.2 b	23.0 b	38 c
Soybean cv. IAC-17	426 a	31.9 a	13.3 b	27 c
Sunflower IAC-Uruguai	429 a	4.6 d	92.4 a	-
Sunn hemp IAC 2	449 a	17.2 c	26.1 b	69 b
Velvet bean	446 a	21.6 b	20.7 b	62 b
[1]C. V. %	2.8	19.1	19.6	13.7

Means followed by the same letter in each column are not different (Comparisons among means were made according to Scott-Knott test, p = 0.05).
[1]Coefficient of variation. Adapted from [14].

Table 5. Carbon and nitrogen concentration, carbon to nitrogen ratio, and N derived from biological N$_2$ fixation (BNF) in the aboveground parts of the rotational crops at harvesting.

The rate of natural colonization with AMF was relatively high in all crops (Table 3). Peanut cv. IAC-Caiapo and sunflower cv. IAC-Uruguai, followed by velvet bean, had at least 64% of root infection with AMF. At the same time, sunflower produced the greatest amount of above-ground biomass, followed by C. juncea and velvet bean. Soybean had the highest grain yield (Table 3) and also presented a considerable percentage of root infection with AMF: 56% (Table 3). Besides the symbiotic association with rhizobia, roots of the legumes can be colonized by fungi of the family Endogonaceae that form vesicular-arbuscular (VA) endomycorrhizas, which help enhance the uptake of phosphorus and other nutrients [34].

Results of a nursery study on the effect of a short season pre-cropping with different mycotrophic herbaceous crops on growth of arbuscular mycorrhiza-dependent mandarin orange plants at an early stage after transplantation were presented by [20]. Mandarin orange seedling plants 180 days after transplantation showed variation in shoot growth in response to single season pre-cropping with seven different crops—maize, Paspalum millet, soybean, onion, tomato, mustard, and ginger, and two non-cropped fallow treatments— non-weeded and weeded fallows. Net growth benefit to the orange plants due to the different pre-crops and the non-weeded fallow treatment over the weeded fallow treatment plants showed a highly positive correlation with mycorrhizal root mass of the orange plants as it varied with the pre-crop treatments. Increase in citrus growth varied between 0 and 50% depending upon the mycorrhizal root mass of the pre-crops and weeds, AMF spore number, and infective inoculum density of the pre-cropped soils. These pre-crop variables individually and cumulatively contributed to the highly significant positive correlation between the AMF potential of the pre-cropped soils and growth of mandarin orange plants through their effect on mycorrhizal root mass development (i.e. extent of mycorrhization) of the mandarin orange plants. The choice of a pre-crop from the available options, grown even for a short season, can substantially alter the inherent AMF potential of soils to a significant influence on the performance of the mycorrhiza-dependent orange plant. The relationship between soil mycorrhizal potential left by a pre-crop and mycorrhizal benefit drawn by the succeeding AMF responsive plant can be of advantage for the exploitation of native AMF potential of soils for growth and nutrition management of crops in low nutrient, low input–output systems of production [20].

Sugarcane yield increased more than 30%, in average, due to the rotational crops as compared with the control treatment; those benefits lasted up to the third harvest (Table 6). In the first cutting, sunflower was the rotational crop that induced the greater yield increase, followed by peanut cv. IAC-Caiapó, and soybean cv. IAC 17. [35] observed that sunn hemp residues increased the sugarcane yield; in the first harvest after the green manure, the effect of the legume crop was better than that of chemical fertilization with nitrogen. Similar results were reported later by [36], with a yield rise of 15.4 tons ha^{-1} of sugarcane stalks, which represented about 24% increase in relation to the control. Positive effects on stalk yields were also found by [31] when sugarcane was grown after *Crotalaria spectabilis*, and by [36], who cultivated sugarcane after sunn hemp and velvet bean.

Sunflower was the best rotational treatment, causing an yield increased of around 46% in the first harvest after the rotational crops (Table 6). Meanwhile, in the average of three cuttings, peanut showed an yield increase of around 22% whereas sunflower presented a 10% yield increase; these results are in agreement with those of [31, 36].

The sugar content of sugarcane stalks is important because the raw material remuneration takes into account this parameter. Some crops that preceded sugarcane had a high effect on sugar yield (Table 7); this was observed mainly in the first harvest in areas where sunflower, peanuts and C. juncea were previously cultivated (Table 2). The 3-year average data showed a sugar yield increase, in the best treatment, of 3 t ha^{-1} in relation to the control. These results were already observed by [35, 31] who found an average increase of 2.98 ton^{-1} ha due to green manure crops grown before sugarcane.

Rotational crops	Stem yield			
	First cut	Second cut	Third cut	Average
	---------------------------- ton ha^{-1} ----------------------------			
Control	47.6 Bc	111.2 Aa	50.7 Ba	69.8
Mung bean cv. M146	61.6 Bb	131.9 Aa	54.7 Ba	82.7
Peanut cv. IAC-Caiapó	67.6 Ba	130.6 Aa	58.0 Ba	85.4
Peanut cv. IAC-Tatu	60.6 Bb	114.9 Aa	66.8 Ba	80.8
Soybean cv. IAC-17	67.5 Ba	124.9 Aa	56.7 Ca	83.1
Sunflower cv. IAC-Uruguai	69.5 Ba	105.2 Aa	55.3 Ca	76.7
Sunn hemp IAC 1	65.9 Bab	125.8 Aa	51.1 Ca	80.9
Velvet bean	61.3 Bb	116.3 Aa	61.2 Ba	79.6
Average	62.7	120.1	56.8	
SEM[1]	0.85	3.80	1.65	

Means followed by the same lower-case letter in the columns and capital letter in the rows are not different (Comparisons among means were made according to Tukey-Kramer test, p > 0.1).
[1]Standard error of the mean. SEM for comparison of rotational crops is 4.22. Adapted from [14].

Table 6. Yield of millable stems of sugarcane grown after rotational crops planted before the first sugarcane cycle.

Rotational crop	Sugar yield[1]				
	First cut	Second cut	Third cut	Average	SEM[2]
	------------------------------------- ton ha^{-1} -----------------------------------				
Control	6.9 Bb	18.1 Aa	7.5 Ba	10.3	1.4
Mung bean cv. M146	9.3 Ba	19.6 Aa	8.3 Ba	12.4	1.7
Peanut cv. IAC-Caiapó	9.9 Ba	21.2 Aa	8.9 Ba	13.3	1.6
Peanut cv. IAC-Tatu	8.8 Cab	18.5 Aa	10.5 Ba	12.6	1.3
Soybean cv. IAC-17	10.0 Ba	17.7 Aa	8.8 Ba	12.2	1.3
Sunflower cv. IAC-Uruguai	10.3 Ba	15.5 Aa	8.1 Ca	11.3	1.0
Sunn hemp IAC 2	9.3 Ba	19.2 Aa	7.5 Ca	12.0	1.5
Velvet bean	9.2 Ba	18.5 Aa	9.5 Ba	12.4	1.3
Average	9.2	18.6	8.6		
SEM[2]	0.2	0.6	0.3		

Means followed by the same lower-case letter in the columns and capital letter in the rows are not different (Comparisons among means were made according to Tukey-Kramer test, p > 0.1).
[1] Apparent sucrose content in the cane juice.
[2]Standard error of the mean. Adapted from [14].

Table 7. Sugar yields of three consecutive cuttings of sugarcane grown after rotational crops.

Studying crop rotation with legume plants in comparison with a control with and without a mineral N addition, [35] observed that, after a crop rotation, the sugarcane yield was higher after C. juncea and velvet bean, with 3.0 and 3.2 stalk tons ha^{-1} increase, respectively. The treatments with an addition of N fertilizer but no-rotation with green manure resulted in only 1.1 tons ha^{-1} of a sugar yield increase, in the average of three years, suggesting that the beneficial influence of leguminous plants is not restricted to the N left by the leguminous plants after harvest.

Farmers must combine the resources of land, labor, management, and capital in order to derive the most profit. Since resources are usually scarce, maximizing returns on each one is important. Crop rotations provide income diversification. If profitability of one crop is reduced because of price variation or some unpredicted reason, income is not as likely to be adversely affected as if the whole farm was planted to this crop, provided that a profit potential exists for each crop in a rotation. This is especially important to the farmer with limited capital.

Some of the general purposes of rotations are to improve or maintain soil fertility, reduce the erosion, reduce the build-up of pests and diseases, best distribute the work load, reduce the risk of weather damage, reduce the reliance on agricultural chemicals, and increase the net profits. Crop rotations have fallen somewhat into disfavor because they require additional planning and management skills, increasing the complexity of farming operations.

Crop rotation can positively affect yield and increase profit (Table 8). Except for peanuts, all other rotational crops contributed to raise the net income. This was true both for the green manures (crotalaria juncea and velvet bean), as for the grain crops (soybean, sunflower and mung bean). Peanuts caused an increase in the sugarcane stalk yields relative to the control, especially in the first harvest (Table 8), but the high cost of production of this grain somewhat cancelled out the benefit of this rotation. However, in many sugarcane regions in São Paulo State peanuts are extensively grown in rotation with sugarcane, probably because in those sites yields are higher and the cost of production, lower. Mung beans are a niche crop. Although it provided a relatively high net return in the present study (Table 8), the risks may be high due to the market restrictions and price fluctuations.

Rotational crop	Gross revenue	Cost of production	Net income
	-------------------- U$ ha^{-1} --------------------		
Control	3,710	3,111	599 b
Mung bean cv. M146	6,131	5,118	1,012 a
Peanut cv. IAC-Caiapó	4,784	4,591	193 b
Peanut cv. IAC-Tatu	4,606	4,401	205 b
Soybean cv. IAC-17	4,961	3,624	1,337 a
Sunflower cv. IAC-Uruguai	4,431	3,584	847 a
Sunn hemp IAC 2	4,263	3,195	1,068 a
Velvet bean	4,193	3,212	981 a
C.V.(%)	-	-	24.1

Means followed by the same letter in the column are not different (Comparisons among means were made according to Scott-Knott test, p > 0.05).
[1]Gross revenue includes sales of the three harvests of sugarcane plus grains of rotational crops. Cost of production includes land and crop management, chemicals, feedstock, and harvesting costs of all sugarcane and rotational crops, but excludes land rental. Adapted from [14].

Table 8. Economic balance[1] of sugarcane production including revenues and costs of crop rotation.

4. To evaluated the recovery of nitrogen by sugarcane when applied green manure crop and mineral N

To evaluate the utilization of nitrogen by sugarcane (*Saccharum spp.*) fertilized with sunn hemp (SH) (*Crotalaria juncea* L.) and ammonium sulfate (AS):

The presence of a green manure crop and mineral N applied together caused some soil alterations that could be detected in samples collected in the sugarcane planting and harvesting seasons (Table 9). There was an increase in calcium and magnesium availability, and consequently in base saturation and pH, in relation to the AS-^{15}N treatment, at planting. Similar results were obtained by [38], who worked with four velvet bean cultivars, velvet bean, Georgia velvet bean, cow itch, and cratylia. The presence of green manure caused a significant sum of bases increase, due to increases in calcium and magnesium; consequently, treatments involving velvet bean showed higher CEC values. The presence of organic acids in the plant mass could be the reason for this change.

During sugarcane harvest, increases in Mg concentration, pH, and base saturation (V%) were observed in the treatments containing SH-[15]N+ AS in relation to the treatment containing AS-^{15}N alone. Also, a significant reduction in potential acidity was observed in treatments containing SH-[15]N+ AS in relation to the treatment containing AS-^{15}N alone (Table 9).

Treatment	Soil sampling at sugarcane planting					
	pH (CaCl2)	Ca	Mg	H+Al	SB	V
	0.01mol l[-1]	------------------mmolc dm[-3]------------------				----- % -----
Control	5.1 ab	20.5 ab	14.5 ab	37.8 a	35.4 ab	48.2 a
AS-[15]N[2]	4.7 b	15.8 b	9.8 b	47.0 a	25.9 b	36.0 a
SH + AS -[15]N	5.3 a	24.8 a	17.8 a	32.0 a	42.8 a	55.8 a
SH-[15]N	5.0 ab	18.0 ab	13.0 ab	39.0 a	31.4 ab	44.5 a
Mean	5.0	19.8	13.8	39.0	33.9	46.1
C.V.%	5.12	7.55	10.76	20.77	6.81	22.52
	Soil sampling at sugarcane harvest					
Control	5.0 ab	17.8 a	14.0 ab	39.8 ab	32.2 a	44.5 ab
AS-[15]N	4.7 b	15.3 a	9.8 b	46.5 a	25.4 a	35.8 b
SH + AS -[15]N	5.6 a	24.7 a	26.8 a	25.5 b	44.4 a	67.5 a
SH-[15]N	5.0 ab	19.0 a	15.3 ab	36.3 ab	34.5 a	48.5 ab
Mean	5.0	18.0	16.4	37.0	33.5	49.1
C.V.%	8.00	30.88	15.18	25.87	32.45	29.57

Means followed by different letters in columns in each sampling season are different (Comparisons among means were made according to Tukey test P < 0.05).
[2] Treatments were: Control (no N fertilizer applied), AS-[15]N (^{15}N-labeled ammonium sulfate), SH + AS -[15]N (Sunn hemp + ^{15}N-labeled ammonium sulfate), SH-[15]N (^{15}N-labeled Sunn hemp). Adapted from [12].

Table 9. Chemical characterization of the soil (0.0-0.2 m depth) in the sugarcane planting and harvesting seasons.

The presence of organic acids in decomposing plant residues can help Mg movement in the soil [39]. Crops with high C:N ratio may release N more slowly and cause an increase in N uptake by succeeding crop In addition, rotational plants that were grown before sugarcane could recycle nutrients that would otherwise be leached contribute with N derived from BNF and keep some elements in plant available forms, which could be transformed into more recalcitrant forms if the soil lie fallow for some time.

There was no variation in nutrient contents for macronutrients N and P, and for micronutrients B and Zn in sugarcane stalks at harvest time (Table 10). However, there were differences in Ca and K contents; the latter showed higher values in treatments involving fertilizer application, either mineral or organic, while Ca showed a higher value in the treatment with green manure and mineral N, indicating better nutrition with this element in the treatment containing higher nitrogen supply.

Nitrogen and potassium absorption is greatly influenced by moisture; this relation has been known for a long time [40], and the fact that treatments involving green manure crops maintained environments with higher moisture due to soil mulching with plant mass could have favored better potassium nutrition. With regard to calcium, nitrogen seems to favor absorption [41].

Treatment	N	K	P	Ca	Zn	B
	Contents determined in sugarcane stalks at harvest					
	---------------g kg^{-1}----------------				--mg kg^{-1}--	
Control	7.2 a	3.3 b	0.8 a	1.6 b	10.9 a	12.1 a
AS-^{15}N^2	8.1 a	6.7 a	0.9 a	1.7 b	15.3 a	14.9 a
SH-^{15}N	7.7 a	7.1 a	0.9 a	1.8 b	13.3 a	14.8 a
SH + AS -^{15}N	8.8 a	8.5 a	1.0 a	2.4 a	13.7 a	15.4 a
Mean	8.0	6.4	0.9	1.88	13.3	14.3
C.V.%	11.52	27.89	15.61	8.14	19.80	18.00

Means followed by different letters in columns are different (Comparisons among means were made according to Tukey test P < 0.05).
[2] Treatments were:Control (no N fertilizer applied), AS-^{15}N (^{15}N-labeled ammonium sulfate), SH-^{15}N (^{15}N-labeled Sunn hemp), SH + AS -^{15}N (Sunn hemp + ^{15}N-labeled ammonium sulfate). Adapted from [12].

Table 10. N, K, P, Ca, Zn, and B contents in sugarcane stalks at harvest time.

When sugarcane was cultivated for five years and was harvested three times. ^{15}N recovery was evaluated in the two first harvests. In the sum of the three harvests, the highest stalk yields were obtained with a combination of green manure and inorganic N fertilizer; however, in the second cutting the yields were higher where sunn hemp (SH) was used than in plots with ammonium sulfate (AS) (Table 11).

Millable stalk yields of the first cycle (plant cane, harvested 18 months after planting) were higher than those of the second and the third cycle (Table 11). The yield decline with time is common, especially in the cases such as the present experiment when only the first cycle crop was fertilized in order to evaluate the residual effect of N application in the mineral or green manure forms. In the first year the stalk yield was numerically higher in plots fertilized with a combination of green manure and AS; however, in the second year the plots that received SH produced more cane than those fertilized only with AS or the control treatment, indicating that the green manure applied before planting still affected plant growth and yield after 34 months. In the third cycle, there were no differences among the treatments, showing that the residual effect of both N sources had disappeared (Table 11). In the sum of three cuttings, the combination of AS and green manure resulted in highest yields.

Treatments[2]	Harvests			Total of three cuttings	Mean ± SEM[3]
	24 Aug 2002	08 Oct 2003	20 Sep 2004		
	------------------------- Stalk yield, Mg ha^{-1} -------------------------				
Control	86.0 Ba	61.1 Bab	47.1 Ab	b194.2	64.7 ± 4.6
AS^{15}N	106.2 ABa	64.7 Bb	42.3 Ab	ab213.2	71.1 ± 4.6
SH + AS^{15}N	128.7 Aa	84.5Ab	45.0 Ac	a258.2	86.1 ± 4.6
SH^{15}N	92.4 ABa	83.8 Aa	41.2 Ab	ab217.3	72.4 ± 4.6
Mean ± SEM	103.3 ± 3.8 a	73.5 ± 3.8 b	43.9 ± 3.8 c	215.4 ± 18.9	
	------------------------POL, Mg ha^{-1} -------------------------				
Control	11.9	10.4	17.9	b40.2	13.5 ± 0.7 B
AS^{15}N	14.9	11.1	17.5	ab43.5	14.5 ± 0.6 AB
SH + AS^{15}N	17.0	14.1	18.4	a49.5	16.5 ± 0.6 A
SH^{15}N	12.9	14.2	18.1	ab45.2	15.1 ± 0.6 AB
Mean ± SEM	14.2 ± 0.9 b	12.4 ± 0.9 b	18.0 ± 0.9 a	43.8 ± 2.4	

Means followed by a different letter lower-case letter, in the rows, and upper-case letter, in the columns, are different (Comparisons among means were made according to Tukey-Kramer p ≤ 0.1). Means followed by superscript letters differ vertically (Comparisons among means were made according to Tukey-Kramer p ≤ 0.1).
[1] Cane was planted on 01 Mar 2001. POL =apparent sucrose content in the cane juice.
[2] Treatments were: Control (no N fertilizer applied), AS^{15}N (^{15}N-labeled ammonium sulfate); SH + AS^{15}N (Sunn hemp + ^{15}N-labeled ammonium sulfate); SH^{15}N (^{15}N-labeled Sunn hemp).
[3] Standard error of the mean. Adapted from [17].

Table 11. Millable stalk yield and POL[1] of sugarcane plants in three consecutive harvests as a function of N applied at planting as ammonium sulfate (AS) or Sunn hemp (SH) green manure[1]

[36] showed evidence of the positive effect of green manure fertilization with sunn hemp in sugarcane, with greater sugarcane yield increase than with the application of 40 kg ha^{-1} mineral N to the soil. [43], studying lupine in maize, and [44], studying velvet bean and sunn hemp in rice, could not find response to mineral N applied after green manure, and no N fertilizer was needed when vetch (*Vicia spp.*) was grown after wheat, and when cotton followed faba beans [42].

The effect of fertilizer source on sugar concentration was less evident. In the average of three cuttings, the value of pol in plots, treated with both AS + SH was higher than in that observed in plots that received no N (Table 11). Pol in cane juice was higher in the third cutting than in the two previous ones. Variations in pol measurements among cropping seasons are usually more affected by environmental conditions (temperature and drought) that determine cane maturation than by nutrition. However, high N tends to decrease sugar content and delay maturation [45]; therefore, after two years with no N fertilization, sugar content in cane plants was more likely to be high.

The recovery of N by the first two consecutive harvests accounted for 19 to 21% of the N applied as leguminous green manure and 46 to 49% of the N applied as AS.

Nitrogen derived from AS and SH in the leaves and top parts of the sugarcane plant, excluding the stalks, varied from 6.9 to 12.3 % of the total N at the end of the first cycle (plant cane) and was not affected by N source (Table 12). But the amounts of N from both sources accumulated in the leaves and tops were in the range of only 4.5 to 6.0 kg ha[-1], which represent a recovery of 6.4 to 8.1% of the N applied as AS and 2.7 and 3.1% of the N from the green manure (Table 12). The recovery of ^{15}N in the second cycle decreased when the N source was the inorganic fertilizer. In the second year the percentage of N derived from sunn hemp was greater than that from the AS, indicating a slightly higher residual effect of the green manure (Table 12).

Sampling dates	Treatments[2]				Mean ± SEM[3]
	AS[15]N	SH-[15]N + AS	SH[15]N	AS[15]N+ SH	
	-------------------------------- Ndff ,% -------------------------				
24 Aug 2002	12.3 Aa	11.1 Aa	10.9 Aa	6.9 Aa	10.3 ± 1.1
08 Oct 2003	1.7 Bb	5.5 Aa	4.1 Aab	1.7 Bb	3.2 ± 1.1
Mean ± SEM	7.0 ± 1.6	8.3 ± 1.6	7.5 ± 1.6	4.3 ± 1.6	
	--------------------------- QNdff, kg ha[-1] ----------------------				
24 Aug 2002	5.7	6.0	5.2	4.5	5.4 ± 0.6 A
08 Oct 2003	1.8	6.8	4.6	2.9	4.0 ± 0.6 A
Mean ± SEM	3.7 ± 1.0 a	6.4 ± 1.0 a	4.9 ± 1.0 a	3.7 ± 1.0 a	
	-------------------------------- R, % ------------------------------				
24 Aug 2002	8.1 Aa	3.1 Aa	2.7 Aa	6.4 Aa	5.1 ± 0.6
08 Oct 2003	2.6 Ba	3.5 Aa	2.3 Aa	4.1 Aa	3.1 ± 0.6
Mean ± SEM	5.3 ± 0.9	3.3 ± 0.9	2.5 ± 0.9	5.3 ± 0.9	

Means followed by a different letter lower-case letter, in the rows, and capital letter, in the columns, are different [Comparisons among means were made according to Tukey-Kramer and F′ tests (p ≤ 0.1), respectively].
[1] Cane was planted on 01 Mar 2001.
[2] Treatments were: Control (no N fertilizer applied), AS[15]N (^{15}N-labeled ammonium sulfate); SH + AS[15]N (Sunn hemp ^{15}N-labeled ammonium sulfate); SH[15]N (^{15}N-labeled Sunn hemp).
[3] Standard error of the mean. Adapted from [12].

Table 12. Percentage (Ndff) and quantity (QNdff) of nitrogen in leaves derived from the labeled fertilizer source and nitrogen recovery (R) in samples taken in the first and second harvests[1].

The percentage of N derived from the AS or SH accumulated in the stalks harvested in the first cycle were similar and ranged from 7.0 to 10.5% of the total N content. In the plant cane cycle the amounts of N in the stalks that had been applied as inorganic or organic fertilizers were higher than those measured in the leaves and tops and varied from 27.3 to 24.1 kg N ha^{-1} (Table 13). The recovery of N derived from inorganic fertilizer - 30.1 to 34.4% - was higher than that of the sunn hemp - 8.8 to 9.8%. However, in the second harvest the N the green manure supplied more N to the cane stalk than AS (Table 13). The difference in the amounts of N in the sugarcane plants derived from green manure and mineral fertilizer in the ratoon crop was around 1to 2 kg ha^{-1} N in leaves and tops (Table 12) and 4 to 7 kg N ha^{-1} in the stalks (Table 13), which were relatively small compared to the amounts of N accumulated in the ratoon plants (179 kg N ha^{-1} in plants supplied with AS and 243 kg ha^{-1} N in the SH treatments, or a 64 kg N ha^{-1}difference - Table 13). These results suggest that the effect of green manure on the yield of the second ratoon crop (Table 11) may not be only due to the extra N supply, but rather to other beneficial role of green manure on soil physical-chemical or biological activity properties.

Adding up the amounts of N taken up by the sugarcane plant and contained in the above-ground parts of the plant (leaves, tops and stalks), AS supplied 32.4 to 34.2 kg N ha^{-1}or about 46 to 49% of N recovery; the N taken up by sugarcane from sunn hemp varied from 37.4 to 40.0 kg ha^{-1}, which represented 19.1 to 20.8% N recovery (Tables 12 and 13).

The recovery of N from fertilizers by sugarcane is usually lower than that of grain crops: the latter varies from 50 to 70% [46] whereas for sugarcane the figures vary from 20 to 40% [47-49, 25]. Results of several studies show that the utilization of N from green manure by subsequent crops rarely exceeds 20% [43, 50, 12, 51] and most of the N remains in the soil, incorporated in the organic matter fraction. In the present study the application of AS along with SH increased N utilization by sugarcane plants. This result is in line with that of [44] who used an organic source isolated or combined with an inorganic fertilizer in rice crops and concluded that the green manures improved the mineral N utilization, resulting in N use efficiency of up to 79%.

In a study in pots [51] observed that maize plants took up more N from sunn hemp incorporated to a sandy soil (Paleudalf) than to a clayey soils (Eutrudox) and that the N derived from the roots was more recalcitrant than that of the shoots. Between 50 and 68% of the ^{15}N of the sunn hemp shoots remained in the soil whereas the figures for roots varied from 65 to 80%. Unaccounted for ^{15}N, probably lost in gaseous forms, varied from 5 to 15% of the sunn hemp N [51].

In a detailed account of the first year of the present experiment, [12] showed that 8 months after planting, the recovery by sugarcane plants (above ground parts) of the N derived from AS or from sunn hemp was similar: 3 to 6% of the added N. However, 12 and 15-month-old sugarcane plants recovered between 20 and 35% of the AS but only 6 to 8% of the sunn hemp-derived N.

The percentage of recovery of the inorganic fertilizer N contained in the stalk when the sugarcane plants were harvested after 18 months of planting varied from 30 to 34%; the corresponding figures for the N derived from sunn hemp were significantly lower: around 9 to 10% (Table 13). The residual effect of the N from both sources in the second harvest of the sugarcane plant was similar: between 4 and 6% of the N supplied at planting as AS or SH was recovered in the stalks of the sugarcane plant 31 months after planting (Table 12).

Sampling dates	Treatments[2]				Mean ± SEM[3]
	AS¹⁵N	SH ¹⁵N + AS	SH¹⁵N	AS¹⁵N + SH	
---------------------------- Ndff, % ------------------------					
24 Aug 2002	10.5 Aa	7.0 Aa	8.2 Aa	10.3 Aa	9.0 ± 1.2
08 Oct 2003	1.4 Bb	3.8 Aa	3.7 Aa	1.7 Bb	2.6 ± 0.1
Mean ± SEM	6.0 ± 1.2	5.4 ± 1.2	5.9 ± 1.2	6.0 ± 1.2	
-------------------------- QNdff, kg ha⁻¹ ----------------------					
24 Aug 2002	24.1 Aa	19.3 Aa	17.3 Aa	21.1 Aa	20.4 ± 2.78
08 Oct 2003	2.7 Bb	8.6 Aa	10.3 Aa	3.9 Bb	6.4 ± 0.8
Média ± SEM	13.4 ± 2.6	14.0 ± 2.6	13.8 ± 2.6	12.5 ± 2.6	
--------------------------------- R, % --------------------------------					
24 Aug 2002	34.4 Aa	9.9 Abc	8.8 Ac	30.1 Aab	20.8 ± 1.9
08 Oct 2003	3.9 Ba	4.4 Aa	5.3 Aa	5.6 Ba	4.8 ± 1.9
Mean ± SEM	19.1 ± 3.2	7.1 ± 3.2	7.0 ± 3.2	17.8 ± 3.2	
-------------------- Cumulative N kg ha⁻¹------------------------					
24 Aug 2002	177.4	235.6	257.0	220.4	222.6 ± 9.2 A
08 Oct 2003	181.0	190.8	228.0	270.8	217.6 ± 9.2 A
Mean ± SEM	179.2 ± 12.0 a	213.2 ± 12.0 a	139.9 ± 12.0 a	245.6 ± 12.0 a	

Means followed by a different letter lower-case letter, in the rows, and capital letter, in the columns, are different [Comparisons among means were made according to Tukey-Kramer and F′ tests (p ≤ 0.1), respectively].
[1] Cane was planted on 01 Mar 2001
[2] Treatments were: Control (no N fertilizer applied), AS¹⁵N (¹⁵N-labeled ammonium sulfate); SH + AS¹⁵N (Sunn hemp + ¹⁵N-labeled ammonium sulfate); SH¹⁵N (¹⁵N-labeled Sunn hemp).
[3] Standard error of the mean. Adapted from [12].

Table 13. Percentage (Ndff) and quantity (QNdff) of nitrogen derived from the labeled fertilizer source, nitrogen recovery (R) in sugarcane stalks and nitrogen accumulated in samplings carried out in the first and second harvestings[1].

In the present study, about 69% of the N present in the sunn hemp residues were from BNF. The data obtained in the present study are also in agreement with those obtained by [18] for green manure produced in the field, in the inter-rows of the ratoon crop.

Perin [32] found substantial amounts of N derived from BNF present in the above ground parts of sunn hemp (57.0%) grown isolated and 61.1% when intercropped with millet (Pennisetum glaucum, (L.) R. Brown) (50% seeded with each crop). The sunn hemp+millet treatment grown before a maize crop resulted in higher grain yield than when sunn hemp alone was the preceding rotation. This effect was not observed when N-fertilizer (90 kg N ha⁻¹) was added; Perin [32] concluded that intercropping legume and cereals is a promising biological strategy to increase and keep N into production system under tropical conditions.

No difference was observed in relation to the cumulative N listed in Table 10. The cumulative N results are similar to those found by [47], who obtained, during plant cane harvesting, mean values of 252.3 kg ha⁻¹ cumulative nitrogen, with high nitrogen and plant material accumulation during the last three months, as also observed by [49]. The nitrogen contents found in the above-ground part of sugarcane, Table 10, are in agreement with results of [47].

As the amounts of N applied as AS or SH to sugarcane in the first cycle were different (70 kg N ha^{-1} as AS and 196 kg N ha^{-1} as SH), the quantities of N derived from the green manure in the second harvest were larger than those from the inorganic fertilizer, although the percentage of N recovery was similar (Table 12).

Because less N derived from the green manure was recovered by the sugarcane plant in the first cycle it would be expected that a higher proportion of that N would be taken up in the second cycle (first ratoon), but this did not happen. It seems that the residual N that is incorporated to the soil organic matter has a somewhat long turnover. Other authors have reported low recovery (about 3.5% of the N) by the second crop after sunn hemp cover crop [52] or hairy vetch (*Vicia villosa* Roth) plowed into the soil [53]. Low recovery of residual N has also been observed for inorganic fertilizer sources: less than 3% of the N derived from fertilizers was taken up by soybeans (*Glicine max* (L) Merril) [54], maize (*Zea Mays* L.) [52], [53] or sugarcane (*Saccharum spp*) [55], results similar to those obtained in the present study (Table 12 and 13).

The amounts of inorganic N, derived from both N sources, present in the 0-0.4 m layer of soil in the first season after N application and were below 1 kg ha^{-1}.

The average concentration of inorganic nitrogen in the 0-40 cm layer of soil was relatively low in most samples taken after 8, 12, 15, and 18 months of planting (Table 14). Samples taken in February, in the middle of the rainy and hot season, presented somewhat higher values of (NH_4^+ + NO_3^-)-N probably reflecting higher mineralization of soil organic N (Figure 2). Later in the growing season (samples of May and Aug 2002) soil inorganic N content decreased again. This coincides with beginning of the dry season with mild temperatures, when the sugarcane plant reached maturity and probably had already depleted the soil for most of the available N.

Treatments[1]	Sampling dates				
	29 Oct 2001	20 Feb 2002	28 May 2002	24 Aug 2002	Mean ± SEM[2]
	---------------------------------------mg kg^{-1}---------------------------------------				
Control	2.7 Ab	7.3 ABa	2.3 Ab	2.8 Ab	3.8 ± 0.22
AS^{15}N	2.6 Ab	9.1 Aa	2.2 Ab	3.2 Ab	4.3 ± 0.22
SH^{15}N + AS	2.9 Ab	7.0 ABa	2.7 Ab	3.1 Ab	3.9 ± 0.22
SH^{15}N	2.8 Ab	5.8 Ba	1.5 Bc	2.8 Ab	3.2 ± 0.22
SH + AS^{15}N	2.7 Ab	7.2 ABa	3.1 Ab	2.6 Ab	3.9 ± 0.22
Mean ± SEM	2.7 ± 0.19	7.3 ± 0.19	2.4 ± 0.19	2.9 ± 0.19	

Means followed by a different letter lower-case letter, in the rows, and capital letter, in the columns, are different (Comparisons among means were made according to Tukey-Kramer test p ≤ 0.1).
[1] Treatments were: Control (no N fertilizer applied); AS^{15}N (^{15}N-labeled ammonium sulfate); SH + AS^{15}N (Sunn hemp + ^{15}N-labeled ammonium sulfate); SH^{15}N (^{15}N-labeled Sunn hemp).
[2] Standard error of the mean. Adapted from [12].

Table 14. Soil mineral N (NH_4^+ + NO_3^-) determined in four sampling dates during the plant cane cycle. Data are average of samplings of the 0-0.2 and 0.2-0.4 m soil layers.

The percentage of the inorganic N derived from AS or SH present in the soil from the 8th to the 18th month after sugarcane planting represented only 1 to 9% of total inorganic N (Table 15). The proportion of N that was originated from AS decreased with time whereas that from the green manure increased, indicating that the mineralization of this organic source could supply more N at the end of the season (Table 15). Indeed, [12] showed that sugarcane stalks sampled in 15-month old plants had significantly higher percentage of N derived from AS than from SH; in the 18th month that difference had disappeared. Nonetheless, throughout the season, the amounts of inorganic N in the soil derived from either AS or SH were of very little significance for the nutrition of the sugarcane plant – less than 1 kg ha[-1] of inorganic N in a 40 cm soil layer (Table 15), indicating that little residual N is expected in soils grown with this crop. Although the rate of N applied as SH was almost 200 kg N ha[-1], little nitrate leaching losses are expected under the conditions of this experiment.

Treatments[1]	Sampling dates				
	29 Oct 2001	20 Feb 2002	28 May 2002	24 Aug 2002	Mean ± SEM[2]
	---------------------------- Ndff, % ----------------------------------				
AS[15]N	5.9 Aa	0.7 Aa	3.2 Aa	1.0 Ba	2.7 ± 0.57
SH[15]N + AS	2.6 Ab	3.2 Aab	9.0 Aa	5.7 ABab	5.1 ± 0.62
SH[15]N	2.9 Aa	4.3 Aa	7.0 Aa	5.9 Aa	5.0 ± 0.58
SH + AS[15]N	2.9 Aa	0.3 Aa	4.0 Aa	1.3 ABa	2.1 ± 0.62
Mean ± SEM	3.6 ± 0.58	2.1 ± 0.56	5.8 ± 056	3.5 ± 0.55	
	-------------- QNdff, kg ha-1 -------------------------				
AS[15]N					
	0.3 Aa	0.3 Aa	0.4 Aa	0.5 Aa	0.40 ± 0.16
SH[15]N + AS	0.1 Aa	0.2 Aa	0.2 Aa	0.2 Aa	0.18 ± 0.18
SH[15]N	0.1 Aa	0.2 Aa	0.1 Aa	0.2 Aa	0.15 ± 0.16
SH + AS[15]N	0.1 Aa	0.1 Aa	0.1 Aa	0.0 Aa	0.07 ± 0.16
Mean ± SEM	0.15 ± 0.09	0.21 ± 0.09	0.24 ± 0.09	0.22 ± 0.09	

For Ndff: means followed by a different letter lower-case letter, in the rows, and capital letter, in the columns, are different (Comparisons among means were made according to Tukey-Kramer and F tests $p \leq 0.1$), respectively.
For Qndff: means followed by a different letter lower-case letter, in the rows, and capital letter, in the columns, are different (Comparisons among means were made according to Tukey-Kramer test $p \leq 0.1$).
[1] Treatments were: Control (no N fertilizer applied); AS[15]N (^{15}N-labeled ammonium sulfate); SH + AS[15]N (Sunn hemp + ^{15}N-labeled ammonium sulfate); SH[15]N (^{15}N-labeled Sunn hemp).
[2] Standard error of the mean. Adapted from [12].

Table 15. Percent (Ndff) and amount (QNdff) of soil mineral N ($NH_4^+ + NO_3^-$) derived from the labeled fertilizer source (Ndff). Data are average of samplings of the 0-0.2 and 0.2-0.4 m soil layers.

Soil N is often the most limiting element for plant growth and quality. Therefore, green manure may be useful for increasing soil fertility and crop production. With regard to fertilization, organic matter such as a green manure can be potentially important sources of N for crop production [56].

Sugarcane is a fast growing plant that produces high amounts of dry matter. Therefore, it tends to rapidly deplete the soil of inorganic N, especially in soils fertilized with small rates of soluble N as in the case of this study. Cantarella [25] reviewed several Brazilian studies showing little nitrate leaching losses in sugarcane. More recently, [57] showed that only 0.2 kg ha^{-1} NO3$^-$-N derived from 120 kg ha^{-1} of N as urea enriched to 5.04 ^{15}N At% applied to the planting furrow leached below 0.9 m in a sugarcane field, although the total N loss reached 18 kg ha^{-1} N, mostly derived from soil organic matter mineralization or residual N already present in the soil. As in the present study, the data of [57] refer to N applied at the end of the rainy season when excess water percolating through the soil profile is limited (Figure 2).

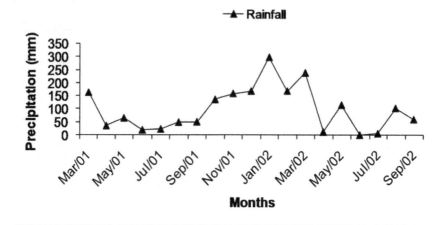

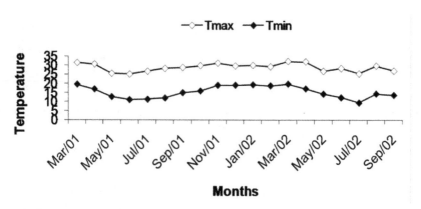

Figure 2. Climatic data for maximum and minimum temperature and rainfall during the first sugarcane growing season (plant cane cycle experiment 2), adapted from [12].

5. To evaluate the effect of biomass on the occurrence of nematodes (*Pratylenchus spp.*) and sugarcane yield after five cuts

The legume most productive was sunn hemp crotalaria juncea IAC-1 with 10,264 kg ha^{-1}, followed by velvet-bean with 4,391 kg ha^{-1} and peanuts IAC-Caiapó and IAC-Tatu with 3,177 kg ha^{-1} and 1,965 kg ha^{-1}, respectively.

There was an increase of Stalk yield of sugarcane in the average of the five cuts, compared to control treatment (Table 16). It can be seen that the effect of planting green manure in the fields of sugarcane promoted reform of benefits in terms of increased productivity of sugarcane, and this is lasting reaching in this case until the fifth cut, and the only treatment that stood out was the rotation of the witness with sunn hemp. Notably, the sunn hemp had higher dry matter production, and this may be a positive influence on growth of sugarcane. After five harvests, sunn hemp crotalaria was the leguminous crop that induced the greatest sugarcane yield, with 30% increase in cane yield and 35% in sugar yield.

Rotational crop	Harvests					
	25 Oct 2001	9 Sep 2002	1Aug 2003	7Nov 2004	6 Oct 2005	Mean
	-------------------------- Stalk yield, Mg ha^{-1} --------------------------					
Sunn hemp IAC 1	145.36	122.30	79.70	51.86	39.30	87.70 A
Velvet bean	141.2	121.88	75.72	51.78	28.12	85.56 AB
Peanut cv. IAC-Tatu	149.92	108.79	74.58	52.16	29.64	83.02 AB
Peanut cv. IAC-Caiapó	122.74	122.30	67.42	49.44	36.78	79.74 AB
Control	129.90	85.31	55.38	46.40	36.15	67.51 B
Mean	138.39 a	113.23 b	71.00 c	50.43 d	34.16 e	

C.V.% (plot) = 7.57, CV% (subplots) = 4.20. Means followed by same lower-case letter in rows and capital letters in columns do not differ (Comparisons among means were made according to Tukey test p> 0.05). For statistical analysis the data were transformed into log (x). Adapted from [16].

Table 16. Millable stalk yield of sugarcane plants in five consecutive harvests as a function of treatments.

Crop rotation with non-host species of nematodes, when well planned, can be an efficient method for integrated control of nematodes. It is common in some areas, the practice of cultivation of Fabaceae in the period between the destruction of ratoon sugarcane field and planting the new. There are several plants used in systems of crop rotation with sugarcane, the most common are crotalarias, velvet beans, soybeans and peanuts. However, depending on nematode species occurring in the area, some of these cultures may significantly increase the population of these parasites. Thus, the sugarcane crop, can be greatly impaired by increasing the inoculum potential of the nematodes [58, 59].

The peanut IAC-Caiapó and velvet bean were the leguminous crops that resulted in the greater percentage of AM fungus. The lowest population of *Pratylenchus spp.* was found in the treatments with peanut IAC-Tatu and IAC-Caiapó (Figure 3).

Thet peanut IAC-Caiapó showed a minimum of 10 nematodes per 10 g of roots and a maximum of 470, while on the control this variation was from 80 to 2,510, indicating the smaller presence of the nematode in treatments with peanut IAC-Caiapó (Figure 3).

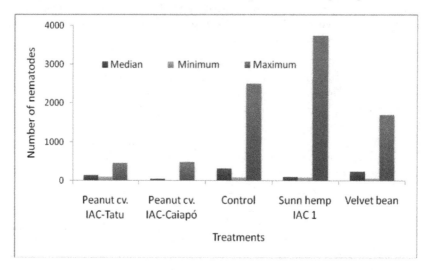

Figure 3. Number of nematodes of the genus *Pratylenchus spp.* by 10 grams of roots of sugar sugarcane cultivation influenced by the previous legume species. Adapted from [16].

6. Conclusions

Crop rotation can positively affect yield and increase profit, contributed to raise the net income. This was true both for the green manures (sunn hemp and velvet bean), as for the grain crops (soybean, sunflower and mung bean). Peanuts caused an increase in the sugarcane stalk yields relative to the control, especially in the first harvest, but the high cost of production of this grain somewhat cancelled out the benefit of this rotation.

However, in many sugarcane regions in São Paulo State (Brazil) peanuts are extensively grown in rotation with sugarcane, probably because in those sites yields are higher and the cost of production, lower. Mung beans are a niche crop. Although it provided a relatively high net return in the present study, the risks may be high due to the market restrictions and price fluctuations.

The biomass of green manure induced a complete N substitution in sugarcane and can cause positively affect yield and increase Ca and Mg contents, sum of bases, pH, and base saturation, and decreasing potential acidity and increase profit.

The combination of inorganic fertilizer and green manure resulted in higher sugarcane yields than either N source separately. The recovery of N from ammonium sulfate was higher in the first year whereas in the green manure presented a longer residual effect and resulted in higher yields of cane in the second cycle. The recovery of [15]N- labeled fertilizers by two successive sugarcane crops summed up 19 to 21% of the N applied as sunn hemp and 46 to 49% of the N applied as ammonium sulfate. Very little inorganic N was present in the 0-40 cm soil layer with both N sources.

The sugar content of sugarcane stalks is important because the raw material remuneration takes into account this parameter in Brazil. Some crops that preceded sugarcane had a high effect on sugar yield; this was observed mainly in the first harvest in areas where sunflower, peanuts, velvet beans and sunn hemp were previously cultivated. The 3-year average data showed a sugar yield increase, in the best treatment, of 3 t ha^{-1} in relation to the control.

The peanut IAC-Caiapó, sunflower and velvet bean were the leguminous crops that resulted in the greater percentage of AM fungus. The lowest population of *Pratylenchus* spp. was found in the treatments with peanut IAC-Tatu and IAC-Caiapó.

After five harvests, sunn hemp crotalaria was the leguminous crop that induced the greatest sugarcane yield, with 30% increase in cane yield and 35% in sugar yield.

Author details

Edmilson José Ambrosano[*], Raquel Castellucci Caruso Sachs
and Juliana Rolim Salomé Teramoto
APTA, Pólo Regional Centro Sul, Piracicaba (SP), Brazil

Fábio Luis Ferreira Dias
APTA, Pólo Regional Centro Sul, Piracicaba (SP), Brazil
Instituto Agronômico (IAC), Campinas (SP), Brazil

Heitor Cantarella
Instituto Agronômico (IAC), Campinas (SP), Brazil

Gláucia Maria Bovi Ambrosano
Universidade Estadual de Campinas, Departamento de Odontologia SocialPiracicaba (SP) Brazil

Eliana Aparecida Schammas
Instituto de Zootecnia, Nova Odessa (SP), Brazil

Fabrício Rossi
Faculdade de Zootecnia e Engenharia de Alimentos (FZEA/USP), Pirassununga, (SP), Brazil

Paulo Cesar Ocheuze Trivelin and Takashi Muraoka
Centro de Energia Nuclear na Agricultura (CENA/USP), Piracicaba (SP), Brazil

Rozario Azcón
Estação Experimental de Zaidin, Granada, Espain

---

[*] Corresponding Author

Acknowledgement

To the technical research support of Gilberto Farias, Benedito Mota, and Maria Aparecida C. de Godoy. To FAPESP and CNPq for the grants. Piraí seeds for green manure and cover crops for the support.

7. References

[1]   Environment Agency, soil: a precious resource, Available:
      http://publications.environment-agency.gov.uk/PDF/GEHO1007BNDB-E-E.pdf   ,
      Accessed 2012 Apr 23, Rio House Waterside Drive, Aztec West Almondsbury, Bristol,
      2007.
[2]   Schipper, L.A. and Sparling, G.P. (2000) Performance of soil condition indicators across
      taxonomic groups and land uses. Soil Sci. Soc. Am. J., 64, 300-311.
[3]   Soil quality for environmental health in http://soilquality.org/home.html Accessed 2012
      Mar 23.
[4]   Pankhurst, C.E., Magarey, R.C., Stirling, G.R., Blair, B.L., Bell, M.J., Garside, A.L., 2003.
      Managementpractices to improve soil health and reduce the effects of detrimental soil
      biota associated with yield decline of sugarcane in Queensland, Australia. Soil Tillage
      Research 72, 125–137.
[5]   Pankhurst, C.E., Stirling, G.R., Magarey, R.C., Blair, B.C., Holt, J.A., Bell, M.J., Garside,
      A.L., 2005. Quantification of the effects of rotation breaks on soil biological properties
      and their impact on yield decline in sugarcane. Soil Biology and Biochemistry 37, 1121–
      1130.
[6]   Shoko, M.D., Zhou, M., 2009. Nematode diversity in a soybean–sugarcane production
      system in a semi-arid region of Zimbabwe. Journal of Entomology and Nematology 1,
      25–28.
[7]   Garside, A.L., Berthelsen, J.E., Richards, C.L., Toovey, L.M., 1996. Fallow legumes on
      the wet tropical coast: some species and management options. In: Proceedings of the
      Australian Society of Sugar Cane Technologists, vol. 18, pp. 202–208.
[8]   Shoko, M.D., Tawira, F., 2007. Benefits of soyabeans as a breakcrop in sugarcane
      production systems in the South Eastern Lowveld of Zimbabwe. Sugar Journal 70, 18–
      22.
[9]   McMahon, G.G., Williams, R.C., McGuire, P.J., 1989. The effects of weed competition on
      ratoon sugar cane yield. In: Proceedings of the Australian Society of Sugar Cane
      Technologists, vol. 10, pp. 88–92.
[10] Cheruiyot, E.K., Mumera, L.M., Nakhone, L.N., Mwonga, S.M., 2003. Effect of
      legumemanaged fallow on weeds and soil nitrogen in following maize (Zea mays L.)
      and wheat (Triticum aestivum L.) crops in the Rift Valley highlands of Kenya.
      Australian Journal of Experimental Agriculture 43, 597–604.
[11] Jannink, J.L.;   Liebman, M.;   Merrick, L.C.   Biomass Production and Nitrogen
      Accumulation in Pea, Oat, and Vetch Green Manure Mixtures. Agronomy Journal 1996.
      88 (2). 231-240.

[12] Ambrosano, E.J.; Trivelin, P.C.O.; Cantarella, H.; Ambrosano, G.M.B.; Schammass, E.A; Guirado, N.; Rossi, F.; Mendes, P.C.D.; Muraoka, T. 2005. Utilization of nitrogen from green manure and mineral fertilizer by sugarcane. Scientia Agricola 62: 534-542.
[13] IBGE. (2010). Systematic Survey of Agricultural Production. Available at: http://www.sidra.ibge.gov.br/bda/default.asp? t=5&z=t&o= 1 & u 1 = 1 & u 2 = 1 & u 3 = 1 & u 4 = 1 & u 5 = 1 & u 6 = 1 & u 7 = 1&u8=1&u9=3&u10=1&u11 =26674&u12= 1&u13=1&u14=1 [Accessed Aug. 16, 2011] (in Portuguese).
[14] Ambrosano, E.J.; Azcón R.; Ambrosano, G.M.B.; Cantarella, H.; Guirado, N.; Muraoka, T.; Rossi, F.; Schammass, E.A.; Trivelin, P.C.O.; Ungaro, M.R.; Teramoto S.J.R. 2010. Crop rotation biomass and arbuscular mycorrhizal fungi effects on sugarcane yield. Scientia Agricola 67: 692-701.
[15] Dinardo-Miranda, L.L.; Fracasso, J.V. 2009. Spatial distribution of plantparasitic nematodes in sugarcane fields. Scientia Agricola 66: 188-194
[16] Ambrosano, E.J.; Ambrosano, G. M. B.; Azcón, R.; Cantarella, H.; Dias F.L.F.; Muraoka, T.; Trivelin, P.C.O.; Rossi, F.; Schammass, E.A.; Sachs R.C.C. Productivity of sugarcane after previous legumes crop. Bragantia v. 70, n. 4, p. 1-9, 2011. (in Portuguese, with abstract in English).
[17] Ambrosano, E.J; Trivelin, P.C.O.; Cantarella, H.; Ambrosano, G.M. B.; Schammass, E.A.; Muraoka,T.; Rossi, F.; 15N-labeled nitrogen from green manure and ammonium sulfate utilization by the sugarcane ratoon. Scientia Agricola, v.68, p.361-368, 2011.
[18] Albuquerque, G.A.C.; Araújo Filho, J.T.; Marinho, M.L. 1980. Green manure and its economic importance. Boletim IAA/PLANALSUCAR 1: 1-10. (in Portuguese).
[19] Azcón, R.; Rubro, R.; Barea, J.M. 1991. Selective interactions between different species of mycorrhizal fungi and Rhizobium meliloti strains and theirs effects on growth, N2 fixation (15N) and nutrition of Medicago sativa L. New Phytologist 117: 399- 404.
[20] Panja, B.N.; Chaudhuri S. 2004. Exploitation of soil arbuscular mycorrhizal potential for AM-dependent mandarin orange plants by pre-cropping with mycotrophic crops. Applied Soil Ecology 26: 249-255.
[21] Philips, J.M.; Hayman, D.S. 1970. Improved procedures for cleaning roots and staining parasitic and vesicular arbuscular mycorrhizal fungi for rapid assessment of infection. Transactions of the British Mycological Society 55: 158-162.
[22] Giovanetti, M.; Mosse, B. 1980. An evaluction of techniques for measuring vesicular arbuscular mycorrhizal spores. New Phytologist 84: 489-500.
[23] Tanimoto, T. 1964. The press method of cane analysis. Hawaiian Planter's Record 57: 133-150.
[24] Ambrosano, E.J.; Trivelin, P.C.O.; Cantarella, H.; Rossetto, R.; Muraoka, T.; Bendassolli, J.A.; Ambrosano, G.M.B.; Tamiso, L.G.; Vieira, F.C.; Prada Neto, I. 2003. Nitrogen-15 labeling of Crotalaria juncea Green Manure. Scientia Agricola 60: 181-184. 67: 692-701.
[25] Cantarella, H.; Trivelin, P.C.O.; Vitti, A.C. 2007. Nitrogen and sulfur in the sugar cane. p. 355-412. In: Yamada, T.; Abdalla, S.R.S.; Vitti, G.C. (ed.) Nitrogen and sulfur in the Brazilian agriculture. Piracicaba: Internacional Plant Nutrition Institute. 722p.

[26] Trivelin, P.C.O.; Lara Cabezas, W.A.R.; Victoria, R.L.; Reichardt, K. 1994. Evaluation of a 15N plot design for estimating plant recovery of fertilizer nitrogen applied to sugar cane. Scientia Agricola.51: 226-234.

[27] Shearer, W.B.; Kohl, D.H. 1986. N2-fixation in field settings: estimations based on natural 15N abundance. Australian Journal of Plant Physiology 13: 699-756.

[28] Bataglia, O.C.; Furlani, A.M.C.; Teixeira; J.P.F.; Furlani, P.R.; Gallo, J.R. 1983. Methods for Chemical Analysis of Plants. Instituto Agronômico, Campinas, SP, Brazil. (in Portuguese).

[29] Silveira, P.M.; Braz, A.J.B.P.; Kliemann, H.J.; Zimmermann, F.J.P. 2005. Accumulation of nutrients in the leaves of Guandu and Estilosantes. Pesquisa Agropecuária Tropical 35: 133-138. (in Portuguese, with abstract in English).

[30] Porto, W.S.; Carvalho, C.G.P.; Pinto, R.J.B.; Oliveira, M.F.; Oliveira, A.C.B. 2008. Evaluation of sunflower cultivars for central Brazil. Scientia Agricola 65: 139-144.

[31] Caceres, N.T.; Alcarde, J.C. 1995. Green manuring with leguminous in rotation with sugar cane (Saccharum ssp). STAB 13: 16-20. (in Portuguese).

[32] Perin, A.; Santos, R.H.S.; Urquiaga, S.; Guerra, J.G.M.; Cecon, P.R. 2006. Sunnhemp and millet as green manure for tropical maize production. Scientia Agricola 63: 453-459.

[33] Guimarães, A.P.; Morais, R.F.; Urquiaga, S.; Boddey, R.M.; Alves, B.J.R. 2008. Bradyrhizobium strain and the 15N natural abundance quantification of biological N2 fixation in soybean. Scientia Agricola 65: 516-524.

[34] Azcón-G. de Aguilar, C.; Azcón, R.; Barea, J.M. 1979. Endomycorrhizal fungi and Rhizobium as biological fertilisers for Medicago sativa in normal cultivation. Nature 279: 325–327.

[35] Wutke, A.C.P.; Alvarez, R. 1968. Restoration of soil for the cultivation of sugar cane.Bragantia 27: 201-217. (in Portuguese, with abstract in English).

[36] Mascarenhas, H.A.A.; Tanaka, R.T.; Costa, A.A.; Rosa, F.V.; Costa, V.F. 1994. Residual Effects of Legumes on the Physical and Economic Yield of Sugarcane, Instituto Agronômico, Campinas, SP, Brazil. (in Portuguese).

[37] Mascarenhas, H.A.A.; Nogueira, S.S.S.; Tanaka, R.T.; Martins, Q.A.C.; Carmello, Q.A.C. 1998. Effect of productivity of crop rotation and summer and sunn henp in the winter. Scientia Agricola 55: 534-537 (in Portuguese, with abstract in English).

[38] Sakai, R.H.; Ambrosano, E.J.; Guirado, N.; Rossi, F.; Mendes, P.C.D.; Cantarella, H.; Arevalo, R.A.; Ambrosano, G.M.B. 2007. Agronomic evaluation of four species of mucuna used as green manure in agroecological systems. Revista Brasileira de Agroecologia 2: 910-913. (in Portuguese, with abstract in English).

[39] Franchini, J. C.; Meda, A. R.; Cassiolato, M. E.; Miyazawa, M.; Pavan, M.A. 2001. Plant residue extracts potential for lime mobility in the soil using a biological method. Scientia Agricola 58: 357-360.

[40] Humbert, R.P.. In: The Growing of sugar cane. Amsterdan: Elsevier, 1968. p.133-309. Nutrition of sugar cane.

[41] Silva, L.C.F.; Casagrande, J.C. Nutrição mineral da cana-de-açúcar (macronutrientes) In: Orlando FilhO, J. Nutrição e adubação da cana-de-açúcar no Brasil. Piracicaba: IAA Planalsucar, 1983. p.77-99. (in Portuguese).

[42] Rochester I. and Peoples M. 2005. Growing Vetches (Vicia Villosa Roth) In Irrigated Cotton Systems: Inputs Of Fixed N, N Fertilizer Savings And Cotton Productivity. Plant Soil. 271: 251-264.

[43] KanthacK, R.A.D.; Mascarenhas, H.A.A.; Castro, O.M.; Tanaka, R.T. Nitrogênio aplicado em cobertura no milho após tremoço. Pesquisa Agropecuária Brasileira, v.26, p.99-104, 1991. (in Portuguese, with abstract in English).

[44] Muraoka, T.; Ambrosano, E.J.; Zapata, F.; Bortoletto, N.; Martins, A.L.M.; Trivelin, P.C.O.; Boaretto, A.E.; Scivittaro, W.B. Eficiência de abonos verdes (crotalária y mucuna) y urea, aplicados solo o juntamente, como fuentes de N para el cultivo de arroz, Terra, v.20, p.17-23, 2002. (in Portuguese, with abstract in English).

[45] Silveira, J.A.G.; Crocomo, O.J. 1990. Nitrogen Assimilation in Sugarcane Plants as Affected by High Levels of N and Vinasse in the Soil. Brazilian Journal of Plant Physiology, 2: 7-15.

[46] Freney, J.R.; Denmead, O.T.; wood., a.w.; saffigna, p.g.; chapman, l.s.; ham, g.j.; hurney, a.p.; stewart, r.l. Factors controlling ammonia loss from trash covered sugarcane fields fertilized with urea. Fertilizer Resarch, 31:341-349, 1992.

[47] Gava, G.J.C.; Trivelin, P.C.O.; Vitti, A.C.; Oliveira, M.W. 2003. Recovery of nitrogen (15N) from urea and cane trash by sugar cane ratoon (Saccharum spp.). Pesquisa Agropecuária Brasileira, 27: 621-630. (in Portuguese, with abstract in English).

[48] Trivelin, P.C.O.; Victoria, R.L.; Rodrigues, J.C.S. Aproveitamento por soqueira de cana-de-açúcar de final de safra do nitrogênio da aquamônia-15N e Uréia – 15N aplicado ao solo em complemento à vinhaça. Pesquisa Agropecuária Brasileira, v.30, p.1375-1385, 1995. (in Portuguese, with abstract in English).

[49] Trivelin, P.C.O.; Rodrigues, J.C.S.; Victoria, R.L. Utilização por soqueira de cana-de-açúcar de inicio de safra do nitrogênio da aquamônia-15N e Uréia – 15N aplicado ao solo em complemento à vinhaça. Pesquisa Agropecuária Brasileira, v. 31, p.89-99, 1996. (in Portuguese, with abstract in English).

[50] Silva, E.C.; Muraoka, T.; Buzetti, S.; Guimarães, G.L.; Trivelin, P.C.O.; Veloso, M.E.C. Utilização do nitrogênio (15N) residual de coberturas de solo e da uréia pela cultura do milho. Revista Brasileira de Ciência do Solo, 30:965-974, 2006. (in Portuguese, with abstract in English).

[51] Ambrosano, E.J.; Trivelin, P.C.O.; Cantarella, H.; Ambrosano, G.M.B.; Schammass, E.A.; Muraoka, T.; Guirado, N.; Rossi, F. 2009. Nitrogen supply to maize from sunn hemp and velvet bean green manures. Scientia Agricola 66: 386-394.

[52] Silva, E.C.; Muraoka, Buzetti, S.; Veloso, M.E.C.; Trivelin, P.C.O. Aproveitamento do nitrogênio (15N) da crotalária e do milheto pelo milho sob plantio direto em Latossolo Vermelho de Cerrado. Ciência Rural, 36:739-746, 2006. (in Portuguese, with abstract in English).

[53] Seo, J. Meisinger, J.J.; Lee, H. Recovery of nitrogen-15-labeled hairy veth and fertilizer applied to corn. Agronomy Journal, 98:245-254, 2006.

[54] Boaretto, A.E.; Spolidoria, E.S.; Freitas, J.G.; Trivelin, P.C.O.; Muraoka, T.; Cantarella, H. Fate of 15N-urea applied to wheat-soybean succession crop. Bragantia, 63:265-274, 2004. (in Portuguese, with abstract in English).

[55] Basanta, M.V.; Dourado Neto, D.; Reichardt, K.; Bacchi, O.O.S.; Oliveira, J.C.M. Trivelin, P.C.O. Timm, L.C.; Tominaga, T.T.; Correchel, V.; Cassaro, F.A.M.; Pires, L.F.; Macedo, J.R. Management effects on nitrogen recovery in a sugarcane crop grown in Brazil. Geoderma, 116:235-248, 2003.
[56] Asagi, N.; Ueno H. 2009. Nitrogen dynamics in paddy soil applied with various 15N-labelled green manures Plant Soil. 322:251–262.
[57] Ghiberto, P.J.L. Libardi, P.L.; Brito, A.S.; Trivelin, P.C.O. Leaching of nutrients from a sugarcane crop growing on an Ultisol in Brazil. Agricultural Water Management, 96:1443-1448, 2009
[58] Novaretti, W.R.T. Nematóides em cana-de-açúcar e seu controle. Informe Agropecuário, v.16, p.37-42. 1992. (in Portuguese, with abstract in English).

Chapter 2

Biomass Production in Northern Great Plains of USA – Agronomic Perspective

Qingwu Xue, Guojie Wang and Paul E. Nyren

Additional information is available at the end of the chapter

1. Introduction

The development of biofuel is an important measure to meet America's energy challenges in the future. In the 2007 Energy Independence and Security Act, the U.S. government mandates that 136 billion liters of biofuel will be produced by 2022, of which 60 billion liters will be cellulosic ethanol derived from biomass [1-3]. Currently, ethanol is one of the biofuels that has been developed extensively. In the U.S., initial efforts for ethanol production were focused on fermentation of sugars from grains (especially maize). However, there have been criticisms for ethanol production from maize because of low energy efficiency, high input cost and adverse environmental impacts [4-5]. Biofuels from biomass feedstocks are more attractive because biomass is a domestic, secure and abundant feedstock. There are at least three major benefits for using biofuels. The very first benefit is national energy security. To reduce the reliance of imported oil for transportation, alternative energy options must be developed. Economically, a biofuel industry would create jobs and ensure growing energy supplies to support national and global prosperity. Environmentally, producing and using more biofules will reduce CO_2 emission and slow down the pace of global warming and climate change.

There are several sources of biomass feedstocks in forest and agricultural lands. The agricultural resources for biomass include annual crop residues, perennial crops, and miscellaneous process residues and manure [2, 3, 6]. Among the agricultural sources, the dedicated biofuel crops based on perennial species have been considered to the future of the biofuel industry and are the focus of intense research [2, 3, 6-8]. In addition, perennial biofuel crops also can provide other environmental and ecological benefits such as improving soil health, providing wild life habitat, increasing carbon sequestration, reducing soil erosion and enhancing water conservation [2, 9]. A key factor for meeting the government's goal is the development of biomass feedstocks with high yield as well as ideal quality for conversion to liquid fuels and valuable chemicals [2-3, 6-8,10].

The Northern Great Plains (NGP) of USA has been identified as an important area for biomass production. In particular, North Dakota is ranked first in potential for producing perennial grasses and other dedicated biofuel crops among the 50 states [10]. With about 1.2 million ha of CRP (Conservation Reserve Program) land and over 2.8 million ha of marginal land that are not suitable for cropping, the state has great potential for liquid biofuel production from biomass crops such as perennial grasses [11]. Before the great potential for biofuel production can be realized, questions still remain for developing management practices and their economic and environmental benefits for biofuel crops, such as appropriate species in certain areas, biomass yield potential and quality, harvesting scheduling (e.g., annual vs. biennial harvest), and effects on soil health and carbon sequestration.

In this paper, we review the current research progress for developing perennial biofuel crops in the NGP, primarily based on long-term field studies. We start to briefly discuss the species selections for biofuel crops in the USA and Europe. Then, we focus on development of crop management strategies for high yield as well as ideal quality. Finally, some possible environmental and ecological benefits from perennial biofuel crops are briefly discussed.

2. Appropriate species for biofuel crops

2.1. Ideal biomass crop for biofuels

There are mainly three goals to develop biomass crop for biofuels: (1) maximizing total biomass yield per year; (2) maintaining sustainability while minimizing inputs; (3) maximizing the fuel production per unit of biomass. To achieve the above goals, an ideal biomass crops should have some attributes as followings: high photosynthesis efficiency (e.g., C4 plants), long canopy (green leaf) duration, low inputs, high water-use efficiency, winter hardiness, no known pests and disease, noninvasive, and uses of existing farm equipment [2]. Based on above criteria, perennial forage crops would be ideal candidates for biofuel crops. The primary purpose for growing perennial crops for biomass production is reducing input and maintenance costs. Economically, using perennial species is more cost effective than annual ones, given the current high costs of fertilizers, pesticides (mainly herbicides) and operation fuels, and low values of lands for growing biomass crops.

2.2. Species for potential biofuel crops

Over the years, many species have been or being evaluated for potential of biofuel crops in the USA and Europe, in which the perennial grasses are dominant (Tables 1 and 2) [12]. In the USA, switchgrass was determined as a model species. In Europe, miscanthus, reed canarygrass, giant reed and switchgrass were chosen for more extensive research programs [12]. In addition, legume species and mixture of multi-species also been evaluated as bioenergy crops [5,13].

2.2.1. Switchgrass and miscanthus

Switchgrass and miscanthus are two dominant species reported in literatures for potential biofuel crops. Switchgrass, a C4 perennial grass, has been designated by the U.S. DOE as primary bioenergy crop and has been extensively studied for over two decades. Several reviews have addressed current research and development issues in switchgrass, from biology and agronomy to economics, and from production to policies [6, 14-18]. The attributes of switchgrass for biofuel production included high productivity under a wide range of environments, suitability for marginal and erosive land, relatively low water and nutrient requirements, and positive environmental benefits [17]. For biofuel purpose, switchgrass can be used to produce ethanol [2, 7, 18]. It also can be used as combustion to co-fire with coal in power plant for electricity. Currently, switchgrass production in southern Iowa is mainly used for combustion [19].

Miscanthus is another C4 tall perennial grass originated in East Asia and has been studied extensively throughout the Europe from the Mediterranean to southern Scandinavia [20]. Comparing with other C4 species (such as maize), miscanthus is more cold tolerance and winter hardy in temperate regions of Europe. It also has a low requirement of nitrogen fertilizer and pesticides. In general, miscanthus has a very high biomass yield potential when it is well established. Lewandowski et al. (2000) [20] reported that the irrigated miscanthus yield can be as high as 30 Mg/ha, and yield under rainfed conditions ranged from 10 to 12 Mg/ha. When compared biomass production in US for switchgrass and Europe for miscanthus, the average yield of miscanthus (22 Mg/ha) was twice as much as the average yield of switchgrass (10 Mg/ha), given the similar temperature, nitrogen and water regimes [21]. A side-by-side study in Illinois showed that average biomass yield in miscanthus (30 Mg/ha) can be 3 times as much as switchgrass (10 Mg/ha) [22]. Compared to switchgrass, miscanthus may require higher input costs because it must be established using rhizome cuttings, which delays full production until the second or third year [20, 21]. In Europe, the primary use of miscanthus biomass is for combustion because of the ideal chemical composition [20]. However, little information is known for the conversion of ethanol from miscanthus.

2.2.2. Reed canarygrass and alfalfa

In addition to switchgrass and miscanthus, two other species, reed canarygrass and alfalfa, have also been studied considerably for biofuel crops. Reed canarygrass is a C3 grass commonly used for hay and grazing in temperate agricultural ecosystems, and can yield 8-10 Mg/ha in the Midwest of USA and northern Europe [6, 12]. Similar to switchgrass, reed canarygrass is difficult to establish and normally has a low yield in the seeding year [6].

Alfalfa, one of the oldest forage crops in the world, has traditionally been used as high quality forage. However, alfalfa may also have some values for biofuel feedstock [13]. In an alfalfa biomass energy production system, the forage could be fractionated into stems and leaves. The stems could be processed to generate electricity or biofuel (ethanol), and the leaves could be sold as a supplemental protein feed for livestock. Currently, researchers in Minnesota are conducting experiments to select dual-use alfalfa varieties and developing management systems [13

English name	Scientific name	Photosynthetic pathway	Yields reported Mg DM/ha/year
Crested wheatgrass	*Agropyron desertorum* (Fisch ex Link) Schult.	C3	16.3
Redtop	*Agrostis gigantea* Roth	C3	Not available
Big bluestem	*Andropogon gerardii* Vitman.	C4	6.8-11.9
Smooth bromegrass	*Bromus inermis* Leyss	C3	3.3-6.7
Bermudagrass	*Cynodon dactylon* L.	C4	1.0-1.9
Intermediate wheatgrass	*Elytrigia intermedia* [Host] Nevski.	C3	Not available
Tall wheatgrass	*Elytrigia pontica* [Podp.] Holub.	C3	Not available
Weeping lovegrass	*Eragrostis curvula* (Schrad.) Nees .	C4	6.8-13.7
Tall Fescue	*Festuca arundinacea* Schreb.	C3	3.6-11.0
Switchgrass	*Panicum virgatum* L.	C4	0.9-34.6
Western wheatgrass	*Pascopyrum smithii* (Rydb.) A. Love	C3	Not available
Bahiagrass	*Paspalum notatum* Flugge.	C4	Not available
Napiergrass (elephant grass)	*Pennisetum purpureum* Schum.	C4	22.0-31.0
Reed canary grass	*Phalaris arundinacea* L.	C3	1.6-12.2
Timothy	*Phleum pratense* L.	C3	1.6-6.0
Energy cane	*Saccharum* spp	C4	32.5
Johnsongrass	*Sorghum halepense* (L.) Pers.	C4	14.0-17.0
Eastern gammagrass	*Tripsacum dactyloides* (L.) L.	C4	3.1-8.0

Table 1. The 18 perennial grass species that were screened by the US herbaceous energy crop research program [12].

2.2.3. Others

Compared to the above four widely studied species, many other species for potential biofuel crops are more regional specific and related to local climatic conditions. In the southern region of the U.S., subtropical and tropical grasses such as bermudagrass and napiergrass have been evaluated as biomass crops [6]. In southwestern Quebec, Canada, a short growing season environment, Madakadze et al. (1998) [23] evaluated 22 warm-season grasses in 5 species (sandreed, switchgrass, big bluestem, Indian grass and cordgrass). They found that the most productive entries were cordgrass and several entries of switchgrass. Switchgrass from high latitude tended to produce less biomass. The sandreed showed little potential for forage or biomass production. This study was conducted using space-planted nursery conditions and these data represent individual plant potential. Thereafter, their studies were only focused on switchgrass under solid sward conditions [23-25].

English name	Scientific name	Photosynthetic pathway	Yields reported Mg DM/ha/year
Meadow Foxtail	*Alopecurus pratensis* L.	C3	6-13
Big Bluestem	*Andropogon gerardii* Vitman	C4	8-15
Giant Reed	*Arundo donax* L.	C3	3-37
Cypergras, Galingale	*Cyperus longus* L.	C4	4-19
Cocksfoot grass	*Dactylis glomerata* L.	C3	8-10
Tall Fescue	*Festuca arundinacea* Schreb.	C3	8-14
Raygras	*Lolium* ssp.	C3	9-12
Miscanthus	*Miscanthus* spp.	C4	5-44
Switchgrass	*Panicum virgatum* L.	C4	5-23
Napier Grass	*Pennisetum purpureum* Schum	C4	27
Reed canary grass	*Phalaris arundinacea* L.	C3	7-13
Timothy	*Phleum pratense* L.	C3	9-18
Common Reed	*Phragmites communis* Trin.	C3	9-13
Energy cane	*Saccharum officinarum* L.	C4	27
Giant Cordgrass/	*Spartina cynosuroides* L.	C4	9
Salt Reedgrass			5-20
Prairie Cordgrass	*Spartina pectinata* Bosc.	C4	4-18

Table 2. Perennial grasses grown or tested as energy crops in Europe [12].

3. Biofuel crops in Northern Great Plains (NGP)

3.1. Species and biomass yields

In NGP, species evaluated for biofuel crops include switchgrass, big bluestem, Indian grass, tall wheatgrass, intermediate wheatgrass, wild rye, alfalfa and sweet clover [11, 26-33]. Switchgrass still remains in most of the studies in NGP. In South Dakota, switchgrass has been evaluated under both conventional farmland and CRP land, and the biomass yield ranged from 2 to 11 Mg/ha [28-30]. In North Dakota, cultivars of switchgrass have been tested in western and central areas in small research plots (Dickinson and Mandan) and biomass yield ranged between 2 to 13 Mg/ha, depending on cultivar [26-27]. In another site (Upham), biomass yield of switchgrass ranged from 2.4 to 10.8 Mg/ha [32]. In an on-farm scale trial, switchgrass yield ranged from 4.6 to 9.9 Mg/ha in Streeter and Munich [8, 34].

For selecting species for biofuel crops, switchgrass still has more advantages than any other species. This is because: (1) the species has been studied extensively in the US in last two decades and the germplasm pool is larger than other species; (2) it is a warm season species and has greater water use efficiency and drought resistance; (3) it is native to North America and there are no concerns about the invasiveness; (4) there are many environmental benefits for growing switchgrass.

Switchgrass plot following the 2011 harvest at Central Grasslands Research Extension Center, Streeter, ND. Photography by Rick Bohn.

In addition to species, environmental factors (e.g., precipitation, temperature, soil type etc.) have large effects on yield and quality in biofuel crops. To address the interactions of species and environment, a ten-year long-term study was initiated and established in 2006 to evaluate ten cool and warm season grasses and mixtures across North Dakota [11]. The 10 entries of species and mixtures were shown in Tables 3. These grasses/mixtures were grown in six environments in five locations across North Dakota. Among the five locations, long term growing season precipitation varies from 318 mm at Williston in the west to 431 mm at Carrington in the east. In general, western ND has a semi-arid environment but eastern ND is more humid [11, 35].

Initial biomass yield data indicated Basin and Altai wildrye showed lower biomass yields than either switchgrass or wheatgrass species (Table 4). Tall wheatgrass and intermediate wheatgrass performed well across environments in North Dakota. In contrast, performance of switchgrass was largely related to environment, particularly the seasonal precipitation. For dryland conditions, studies are still needed to address both establishment and persistence of switchgrass in the future.

Harvesting perennial grasses plots in fall 2007, Streeter, ND.

Entry	Species/mixtures
1	Switchgrass (Sunburst)
2	Switchgrass (Trailblazer or Dakota)
3	Tall wheatgrass (Alkar)
4	Intermediate wheatgrass (Haymaker)
5	CRP Mix [Intermediate wheatgrass (Haymaker) + Tall wheatgrass (Alkar)]
6	CRP Mix [Intermediate wheatgrass (Haymaker) + Tall wheatgrass (Alkar) + alfalfa + Yellow sweetclover]
7	Switchgrass (Sunburst) + Tall wheatgrass (Alkar)
8	Switchgrass (Sunburst) + Big Bluestem (Sunnyview)
9	Switchgrass (Sunburst) + Altai Wildrye (Mustang)
10	Basin Wildrye (Magnar) + Altai Wildrye (Mustang)

Table 3. Species/mixtures of perennial grasses in ten entries used for biomass study across five locations in North Dakota (names in parenthesis are cultivars) [11].

3.2. Chemical composition

Chemical composition of biomass feedstock affects the efficiency of biofuel production and energy output. The major parts of the chemical composition in the perennial biomass feedstocks are lignocellulose including cellulose, hemicellulose, and lignin; and mineral elements such as ash [3, 36-38]. Biomass may be converted into energy by direct combustion or by producing liquid fuels (mainly ethanol) using different technologies. For converting cellulosic biomass into ethanol, the conversion technologies generally fall into two major categories: biochemical and thermochemical [3, 37, 38]. Biochemical conversion refers to the fermentation of carbohydrates by breakdown of feedstocks. Thermochemical conversion includes the gasification and pyrolysis of biomass into synthetic gas or liquid oil for further fermentation or catalysis. Currently, the U.S. Environmental Protection Agency (USEPA) listed six conversion categories from different companies for ethanol from biomass [3]. Different conversion technologies may require different biomass quality attributes. For ethanol production from biochemical process (fermentation), ideal biomass composition would contain high concentrations of cellulose and hemicellulose but low concentration of lignin [37-38]. While for gasification-fermentation conversion technology, low lignin may not be necessary. For direct combustion and some thermochemical conversion processes, high ash content can reduce the effectiveness and chemical output [3, 37-38].

Entry	Hettinger	Williston-dryland	Williston-irrigated	Minot	Streeter	Carrington
			----------Mg/ha ----------			
1	0.0 c[+]	0.2 c	13.0 ab	5.2 cde	4.0 c	12.1 ab
2	0.0 c	0.7 bc	9.6 cd	2.9 e	4.3 c	13.7 a
3	3.4 a	2.2 a	11.2 bc	10.1 a	7.4 a	10.5 bcd
4	1.8 abc	2.7 a	9.2 cd	7.4 bc	6.0 b	10.1 cd
5	3.4 a	2.5 a	10.1 cd	9.4 ab	7.6 a	9.6 d
6	4.0 a	1.8 ab	8.7 d	8.5 ab	5.8 b	10.3 bcd
7	2.0 abc	2.2 a	12.8 ab	9.4 ab	8.3 a	11.4 bc
8	0.0 c	0.7 bc	11.2 bc	4.7 de	3.6 c	12.1 ab
9	0.0 c	0.7 bc	14.3 a	5.8 cd	3.6 c	11.4 bc
10	0.9 bc	0.7 bc	9.0 d	5.8 cd	3.4 c	9.0 d
Mean	**1.5**	**1.4**	**10.9**	**6.9**	**5.4**	**11.0**
LSD (0.05)	**2.5**	**1.3**	**2.0**	**2.2**	**1.8**	**1.1**

[+]In each column, values followed by the same letter were not significantly different based on LSD test at P=0.05.

Table 4. Biomass yields in ten entries with different species/mixtures of perennial grasses harvested in 2007 at five locations in North Dakota (the species/mixture for each entry is shown in Table 3) [11].

Among the perennial grasses for biofuel production, chemical composition of switchgrass has been investigated in many studies [19, 29-31, 35, 39]. There is little information in the lignocellulose contents in other species such as tall and intermediate wheatgrass when they are harvested at fall as biomass feedstocks because these species have been mainly used as forage. As with yield, biomass composition is affected by genetic and environmental factors as well as by management practices such as nitrogen (N) fertilization and harvest timing. In a study in the southern Iowa, both yield and quality traits were different among 20 switchgrass cultivars. The high yielding cultivars generally had low ash content [19]. In NGP, we reported the chemical composition of the above 10 perennial grasses and mixtures shown in Table 3 in 2007 harvest. The contents of neutral detergent fiber (NDF), acid detergent fiber (ADF), acid detergent lignin (ADL), hemicellulose (HCE), cellulose (CE) and ash were determined. Biomass chemical composition was affected by environment and species/mixtures, and their interaction. Biomass under drier conditions had higher NDF, ADL and HCE contents but lower CE contents. Tall and intermediate wheatgrass had higher NDF, ADF and CE but lower ash contents than the other species and mixtures. Switchgrass and mixtures had higher HCE. Tall wheatgrass and Sunburst switchgrass had the lowest ADL content. Biomass with higher yield had higher cellulose content but lower ash content. Combining with higher yields, tall and intermediate wheatgrass and switchgrass had optimal chemical compositions for biomass feedstocks production (Table 5) [35]. In another study in NGP, Karki et al. (2011) showed that tall wheatgrass had similar composition to switchgrass and has potential for ethanol production [39].

Entry	NDF	ADF	ADL	HCE	CE	Ash
	-- g/kg --					
1	733.4 bcd[+]	475.1 c	116.0 e	258.4 bcd	359.1 cd	79.2 ab
2	736.8 bcd	468.5 cd	139.1 bc	268.3 ab	329.4 f	81.2 a
3	792.6 a	535.2 a	116.3 e	257.4 bcd	418.9 a	68.8 de
4	753.5 b	507.1 b	154.5 a	246.4 d	352.6 de	71.3 cde
5	753.6 b	503.8 b	145.9 ab	249.8 cd	358.1 d	70.7 cde
6	745.5 bc	518.0 ab	140.4 bc	227.5 e	377.6 bc	69.5 cde
7	781.9 a	515.9 b	121.3 de	266.1 abc	394.6 b	64.3 e
8	736.8 bcd	459.9 cd	132.5 cd	276.9 a	327.4 f	73.9 bcd
9	723.7 cd	456.2 d	124.7 de	267.1 ab	331.5 f	74.8 abcd
10	715.4 d	461.9 cd	124.2 de	253.5 bcd	337.7 ef	76.3 abc
Mean	**747.3**	**490.2**	**131.5**	**257.1**	**358.7**	**73.0**
LSD (0.05)	**23.6**	**18.6**	**12.5**	**16.9**	**18.9**	**7.1**

[+]In each column, values followed by the same letter were not significantly different based on LSD test at P=0.05.

NDF: Neutral detergent fiber; ADF: Acid detergent fiber; ADL: Acid detergent lignin;

HCE: Hemicellulose (NDF-ADF); CE: Cellulose (ADF-ADL).

Table 5. Biomass compositional parameters in different species/mixtures averaged across six environments (the species/mixture for each entry is shown in Table 3) [35].

3.3. Mixture of multiple species

From a long-term sustainability perspective, the reliance on a single species of perennial crops (monoculture) for biomass production may be risky because of less diversity and more chance to prone to certain pests and diseases. Mixture of multiple species may overcome some problems encountered in monoculture crops. In terms of dedicated biofuel crops such as switchgrass and miscanthus, most previous and current studies are focused on monoculture. Little information is known about the mixture of multiple species and their productivity as compared to monoculture.

In ecology studies, the benefits of mixtures of species over monocultures in terms of sustainability and biodiversity have been recognized in both annual and perennial species [6, 40, 41]. For biofuel purpose, specifically, Tilman et al. (2006) argued that the mixtures of different perennial grasses are more stable, more reliable and more productive than monoculture. Also, the mixtures are more environmentally friendly in terms of energy inputs and greenhouse gas emission. From agronomic standpoint, growing mixtures of multiple species in a large farm scale will face challenges such as selecting species, seeding methods, seeds costs, harvesting and so on. In addition, biomass feedstock quality will be an important factor when considering harvesting mixtures.

4. Management strategies for perennial biofuel crops

4.1. Establishment

Many perennial warm season grasses such as switchgrass are difficult to establish [17]. In an on-farm scale study, net energy value of switchgrass is largely determined by the biomass yield in established year [8]. Therefore, improving crop establishment is a very important step to successfully manage biofuel crops. There are many factors affecting establishment of perennial grasses; however, soil moisture and temperature are the most important ones, and many management practices are related to maintenance of adequate moisture and optimum temperature for seedling development and growth.

Seeding rate (pure live seeds): Typically recommended seeding rate in the US is 4-10 kg/ha for switchgrass based on the review of Parrish and Fike (2005) [17]. Sedivec et al. (2001) provided a detail recommendation for grass varieties for ND, ranging from 2 to 24 lb/ac, depending on species or varieties [42].

Seeding depth: The seeding depth may vary with soil types. However, seeding depth of grasses is generally shallower than cereal crops. For switchgrass in NE, seeding depth ranged from 1.5 cm to 3.0 cm in silt loam soil [43]. In SD, Nyoka et al. (2007) recommended not seeding deeper than 2.5 cm regardless of soil type [44].

Seeding date: Seeding date is largely determined by soil temperature and moisture. For warm season grasses, the ideal temperature for seed germination is between 20-30 oC if no dormancy [44, 46]. In SD, the recommended seeding date is early May to mid-June [44]. In VA, the planting date for switchgrass is much later than for corn but similar to that for millet or sorghum-sudangrass. In conventionally prepared seedbeds, June 1-15 was recommended [47]. In NE, study showed that planting switchgrass in mid-March can significantly increase seedling size as compared to late April and May [48]. Under NGP conditions, early seeding may provide benefit in terms of adequate soil moisture [48]. However, low soil temperature may be a factor for limiting germination and emergence of warm season grasses.

Timing an appropriate seeding date is also important for weed control. In a study conducted in Mississippi, Holmberg and Baldwin (2006) seeded switchgrass monthly from April to October and found that the months with minimum weed biomass were April and June. In addition, rainfall is also a very important factor for determining weed suppression for seeding switchgrass [49].

Seeding methods: switchgrass and other warm season grasses can be seeded under both conventional and no-till conditions. The ideal condition for conventional seeding should be a smooth, firm, clod-free soil for optimum seed placement with drills or culti-packer seeders [44]. The seedbed should be firm enough for good seed-soil contact and a consistent seeding depth [44, 47]. Since switchgrass requires warm weather for seeding, water loss during tillage could be a problem under dry and warm days. As a result, conventional seeding may not be ideal [47].

No-till helps to conserve soil moisture, requires less time and fuel, and eliminates the soil crusting frequently encountered in conventional seedbed [47]. In the literature, the results of comparison of conventional and no-till planting for warm season grass establishment are controversial. However, no-till planting frequently showed advantages over conventional tillage, in terms of soil and water conservation [17].

The warm season grasses can be seeded by drilling as well as broadcasting. For broadcasting method, cultipacking or rolling the seedbed after broadcasting is required to ensure that seeds are sufficiently covered by soil and to improve seed-to-soil contact [44].

Seed size (seed mass): Seed size varies considerably within cultivars as well as seedlots of a single cultivar [50]. In general, seed size is linearly related to seed mass or weight in many grasses and cereal crops. Large seeds normally have advantage over small seeds for germination and emergence [51], and seedling development [52]. Switchgrass seedlings grown from larger seeds developed adventitious roots more quickly than those from small seeds [52]. Even the seedling size associated to seed size was only evident at early stage [53], Vogel (2000) still suggested that selection of populations with larger seeds may improve seedling establishment in switchgrass [18].

Seedling vigor: Seedling establishment can be quantified by a more general term, seedling vigor. Greater seedling vigor refers to larger seedling size, greater ground cover and higher biomass at early stage. In addition to environmental factors, seedling vigor is believed to the single most important trait controlled by genetic variability in establishment capacity of perennial forage crops. Many researchers have used some measure of seedling vigor as a selection criterion to improve establishment capacity, while others have used more indirect measures, such as seed mass or germination rate [54]. As mentioned in the above, seed size is positively related to seeding vigor. However, other factors are also related to seedling vigor. For example, studies in cereal crops in Australia showed that embryo size significantly contributed to seedling vigor in barley [55]. In spring wheat, high protein content also contributed to seeding vigor.

Others: Application of arbuscular mycorrhizal fungi (AMF) has been shown to be effective for enhancing seedling yield and nutrient uptake in switchgrass [56-58]. Hanson and

Johnson (2005) showed that soil PH affected switchgrass germination and the optimum PH is 6.0 [46].

4.2. Weed control during establishment

Weed competition is often a major cause of establishment failure in grasses [16, 17, 44]. Although the weed species varies from region to region and even between nearby locations, perennial forbs and warm-season grass species provide the most severe competition for warm season crops like switchgrass [17].

Application of herbicides generally provides very effective weed control. In switchgrass and other warm season grasses, atrazine has been used almost universally as both pre- and post-emergence herbicides for improving establishment [17]. However, atrazine is only labeled for roadside and CRP lands in some states, not for large area of switchgrass except for a special use in Iowa [17]. Alternatively, switchgrass was companion-planted with corn or sorghum-sudangrass using atrazine [59-60].

There are several other chemicals showing to be effective for controlling weed during switchgrass establishment. For pre-emergence application, Mitchell and Britton (2000) [61] used metolachlor for control of several warm season annual grasses. Chlorsulfuron and metsulfuron showed some efficacy in switchgrass [62]. For post-emergence application, imazapyr, sulfometuron, quincloric, 2, 4-D and dicamba have been reported or recommend for weeds control in switchgrass and other warm season grasses [17, 44]. Non-selective herbicides (e.g, glyphosate and paraquat) have been used to prepare seedbeds for no-till plantings for establishing grasses. In addition, Buhler et al. (1998) listed a few more herbicides that showed potential to provide selective weed control to improve establishment of perennial warm season grasses [63].

Herbicides are generally effective and largely available in the market. However, many herbicides are not currently registered for perennial crops for biomass production [16, 63, 71]. As a result, weed control during the establishment year can not be solely relied on chemical applications. Other control methods must be adopted to achieve the best weed control. Buhler et al. (1998) reviewed weed management in biofuel crops and provided several non-chemical control options. These options include timing seeding date, tillage and cropping practices, using companion crops and clipping. Ultimately, the best weed management strategy will be the integration of various options [63].

The overall goal of non-chemical options for weed control is to create an environment that favors to crop growth and development but disfavors weeds. A typical example is manipulation of seeding date to minimize the weed competition, by changing the relative emergence of crop and weed. In general, if crops emerged earlier than weeds, they would have advantage to acquire resources. Therefore, seeding crops before the weeds emergence is an effective way to avoid weeds pressure.

Several other management practices have been successfully used to increase crop competitive ability. In western US, Canada and Australia, increasing seeding rate has been an effective measure to suppress wild oat in barley and spring wheat [64-66]. Using large seeds also provided competitive benefit in sparing wheat against wild oat [67. 68]. Choosing cultivars with more competitive ability also provided benefit to weed control during establishment stage.

4.3. Nitrogen (N) management and N use efficiency

Like any other crops, optimizing biomass yield and maintaining quality stands require fertilizer inputs for biofuel crops. Currently, nitrogen (N) remains the primary fertilizer used in biofuel crops; therefore, most studies just consider the N application. Although some perennial species such as swithgrass and miscanthus are tolerant to low soil fertility conditions, studies showed that biomass yield responded to N application [16, 69]. Lemus et al. (2008) used 4 N rates (0, 56, 112 and 224 kg/ha) in switchgrass southern Iowa. They found that N application generally improved the biomass yield but the yield response declined as N level increased [19].

The amount of N fertilizer required for any biofuel crop is a function of several factors including yield potential of the site, cultivar, management practices, soil types, and so on. Therefore, the optimum N rate can vary from place to place. For example, a study in Texas using lowland switchgrass cultivar 'Alamo' determined that the optimum N rate was 168 kg/ha [70]. In another study in CRP land of NGP, however, the N rate of 56 kg/ha was optimum for upland switchgrass cultivars [30]. Gunderson et al. (2008) [71] summarized the response of biomass yield of upland switchgrass cultivars to N fertilizer rate. They showed that switchgrass yield even decreased as N rate was over 100 kg/ha (Figure 1). Among the management practices, perennial grass rotating with legume crops or mixture of grass and legumes may reduce N fertilizer inputs and improve their energy balance [71].

Some perennial grasses (e.g., switchgrass and miscanthus) can recycle N from the aboveground shoots to the crown, rhizome, and root in the fall for use in over-wintering and regrowth in the following spring [72]. This mechanism makes an efficient use and reuse of N by plant. However, there is still little information on when and how much of N recycles among plant organs, and how much the N cycling can contribute to over N balance in biofuel crops [7].

Another factor affecting crop N balance is fertilizer use efficiency and N use efficiency (NUE). Take switchgrass as an example, biomass yield varied considerably (up to 5 fold) at the same N application level (Figure 1). Certainly, N was not used efficiently at low yield level. Therefore, improving both fertilization use efficiency and NUE is very important for increasing biomass yield in biofuel crops. In addition, increased efficiency will ultimately reduce the N inputs.

4.4. Water management and Water Use Efficiency (WUE)

In NGP, soil water deficit occurs very frequently during crop growing season because of the highly variable and uneven distribution of seasonal precipitation. In general, biomass yield of switchgrass increased as the amount of seasonal precipitation increased. However, at a given seasonal precipitation level (e.g., 500-600 mm), switchgrass yield ranged from 2 to 25 Mg/ha (Figure 2) [71], indicating the importance of crop WUE and precipitation use efficiency. Ideally, the figure 2 should be converted to the biomass yield as a function of seasonal evapotranspiration (ET) or transpiration (T), not precipitation because crop yield is more closely related to ET or T. Although most field studies have included precipitation information in NGP, there is no detailed information of crop ET, transpiration and water-use efficiency (WUE). The quantification of ET and WUE in biofuel crops under various environmental conditions and management practices will lead to identify the best management strategies. Because both water and N are critical for crop growth, the interaction of water and N becomes important, particularly under dryland conditions. However, there are very few studies on the interactive effects of N and soil moisture on biomass yield and quality in biofuel crops.

4.5. Harvest management

Proper harvest management is important for biofuel crops for high yields and ideal qualities. The harvest management practices include harvest frequency, timing and stubble height. Currently, most studies for harvest management are focused on switchgrass [29, 69, 73]. Although switchgrass can be harvested in 2 times a year in south part of USA [73, 74], swithcgrass in NGP can only be harvested once a year either after anthesis (summer) or killing frost (fall). For maximizing the biomass yields and chemical compositional attributes for biofuels, harvesting in killing frost is an ideal harvest management [29]. Another harvest practice in the NGP is harvesting every other year (biennial harvest). Comparing annual harvest and biennial harvest, average annual biomass yield is generally lower for biennial harvest. The only benefit for biennial harvest is reducing machine operation cost. However, biennial harvest improved the switchgrass stand health if harvested in summer [29]. The reduction of annual biomass yield in biennial harvest was related to species and mixtures in our long-term field study. The reduction in annual biomass yield due to biennial harvesting ranged between 20 to 50 percent. In general, biomass yield of intermediate wheatgrass reduced the most in biennial harvest. However, there was one dryland site that Sunburst switchgrass + Altai wildrye had higher yield on the biennial harvest [11, 75]. Cutting height during harvest also affect biomass yield in perennial grasses. In general, lower cutting stubble resulted in higher biomass yield than higher cutting [75].

4.6. The role of biofuel crops in cropping systems

Given emerging markets for biofuels and increasing production of biofuel crops, new and improved cropping systems are needed to maintain overall productivity as well as sustainability. Introducing perennial crops to the existing cropping systems will face challenges. Boehmel et al. (2008) [76] studied annual and perennial biofuel cropping systems in Germany. They compared 6 systems: short rotation willow coppice, miscanthus, switchgrass, energy corn and 2 annual crop rotation systems (oilseed rape, winter wheat and triticale). The results showed that perennial biomass systems based on *Miscanthus*, switchgrass, or willows could be as productive as energy corn with lower energy inputs. Energy corn had the best energy yield performance but a relatively high energy input.

Anex et al. (2007) [77] proposed that the development of new biofuel crops and cropping systems, in conjunction with nutrient recycling between field and biorefinery, comprise a key strategy for the sustainable production of biofuels and other commodity chemicals derived from plant biomass. Such systems will allow N nutrient to be recovered and reduce fertilizer inputs.

Currently, little information is known how perennial crops interact with annual crops and their benefit in NGP. Perennials, however, are rarely permanent and some annual cropping or innovative combinations of annual and perennial biofuel crops strategically deployed across the farm landscape and combined into synergistic rotations may be necessary in the future. Combining annual biofuel crops such as corn and sorghum into rotations with perennial biofuel crops may benefit biofuel cropping systems [77].

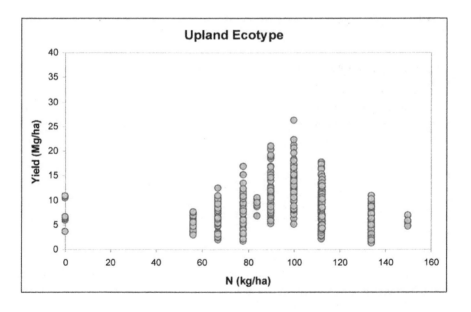

Figure 1. Biomass yield in upland switchgrass as a function of total nitrogen application during the growing season [71].

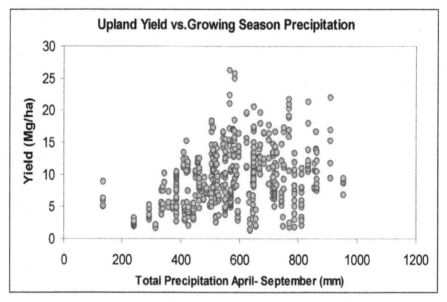

Figure 2. Biomass yield in upland switchgrass as a function of precipitation from April to September [71].

5. Ecological and environmental benefits of biofuel crops

Development of perennial biofuel crops may provide long-term sustainability on these lands by reducing soil erosion, increasing soil organic matter, reducing greenhouse gases and enhancing carbon sequestration [35]. Studies have shown that perennial crops provided many ecological and environmental benefits. Switchgrass and other warm season grasses can be used to control soil erosion, reduce runoff losses of soil nutrients, improve water quality (facilitate the breakdown or removal of soil contaminants), diversify wild life habitats and so on [17, 44]. Roth et al. (2005) [78] showed that proper managing switchgrass harvest can significantly increase grassland birds diversity. More importantly, perennial crops such as switchgrass have been shown to increase carbon sequestration and improve soil quality [9].

The environmental benefits for producing biofuel crops include high energy efficiency and reducing greenhouse gas (GHG) emission. Schmer et al. (2008) [8] evaluated the net energy efficiency and economic feasibility of switchgrass and similar crops in North and Central Great Plains. Switchgrass produced 540% more renewable than nonrenewable energy consumed. Switchgrass monocultures managed for high yield produced 93% more biomass yield and an equivalent estimated NEY than previous estimates from human-made prairies that received low agricultural inputs. Estimated average GHG emissions from cellulosic ethanol derived from switchgrass were 94% lower than estimated GHG from gasoline.

6. Future perspectives for biomass production in the northern great plains

The Northern Great Plains has over 4 million hectares of highly erodible and saline crop land. Development of perennial biofuel crops may provide long-term sustainability on these lands by reducing soil erosion, increasing soil organic matter, reducing greenhouse gases and enhancing carbon sequestration. Although studies are on-going in long-term field experiments, the best management practices are still needed to be developed for producers. The long-term ecological and environmental benefits are also needed to be quantified in the area.

Author details

Qingwu Xue
Texas A&M AgriLife Research and Extension Center at Amarillo, Amarillo, TX, USA

Guojie Wang and Paul E. Nyren
North Dakota State University, Central Grasslands Research Extension Center, Streeter, ND, USA

7. References

[1] United States Congress (2007) Energy Independence and Security Act of 2007, 110th Congress, 1st session, H.R. 6.

[2] U. S. Department of Energy (DOE) (2006) Breaking the Biological Barriers to Cellulosic Ethanol: A Joint Research Agenda, DOE/SC-0095, U. S. Department of Energy Office of Science and Office of Energy Efficiency and Renewable Energy.

[3] United States Environmental Protection Agency (USEPA) (2010). Renewable Fuel Standard Program (RFS2) Regulatory Impact Analysis.
http://www.epa.gov/oms/renewablefuels/420r10006.pdf (Accessed April 2012).

[4] Pimentel D, Doughty R, Carothers C, Lamberson S, Bora N, Lee K (2002) Energy inputs in crop production: comparison of developed and developing countries, *in* Lal, R., Hansen, D., Uphoff, N., and Slack, S., eds., Food Security & Environmental Quality in the Developing World. CRC Press, Boca Raton, FL, p. 129–151.

[5] Tilman, D., J. Hill and C. Lehman. 2006. Carbon-negative biofuels for low-input high-diversity grassland biomass. Science 314: 1598-1600.

[6] Sanderson M.A, Adler P.R (2008) Perennial forages as second generation bioenergy crops. International J. Mol. Sci. 9: 768-788.

[7] Sanderson M.A, Adler P.R, Boateng A.A, Casler M.D, Sarath G (2006) Switchgrass as a biofuels feedstock in the USA. Canadian Journal of Plant Science 86:1315-1325.

[8] Schmer M.R, Vogel K.P, Mitchell R.B, Perrin R.K (2008) Net energy of cellulosic ethanol from switchgrass. Proceedings of the National Academy of Sciences of the United States of America 105: 464-469.

[9] Liebig M.A., Johnson H.A, Hanson J.D, Frank A.B (2005) Soil carbon under switchgrass stands and cultivated cropland. Biomass & Bioenergy 28, 347-354.

[10] Milbrandt A (2005) A Geographic Perspective on the Current Biomass Resource Availability in the United States. *Technical Report, NREL/TP-560-39181*.

[11] Nyren P.E, Eriksmoen E, Bradbury G, Halverson M, Aberle E, Nichols K, Liebig M (2007) The Evaluation of Selected Perennial Grasses for Biofuel Production in Central and Western North Dakota. 2007 Annual Report of Central Grasslands Research Center, NDSU, Streeter.

[12] Lewandowski I, Scurlock J.M.O, Lindvall E, Christou M (2003) The development and current status of perennial rhizomatous grasses as energy crops in the US and Europe. Biomass and Bioenergy. 25: 335-361.

[13] Lamb J.F.S, Jung H.G, Sheaffer C.C, Samac D.A (2007) Alfalfa leaf protein and stem cell wall polysaccharide yields under hay and biomass management systems. Crop Sci. 47: 1407-1415.

[14] Hoekman S. K (2009) Biofuels in the U.S. – challenges and opportunities. Renewable Energy 34: 14-22.

[15] McLaughlin S.B., Kiniry J.R, Taliaferro C.M, Ugarte D.D (2006) Projecting yield and utilization potential of switchgrass as an energy crop. Adv. Agron. 90: 267-297.

[16] Mitchell R, Vogel K.P, Sarath G (2008) Managing and enhancing switchgrass as a bioenrgy feedstock. Biofuels, Bioprod. Bioref. 2: 530-539.

[17] Parrish D.J, Fike J.H (2005) The biology and agronomy of switchgrass for biofuels. Critical Rev. in Plant Sc. 24: 423-459.

[18] Vogel K.P (2000) Improving warm-season forage grasses using selection, breeding, and biotechnology. p. 83–106 *In* K.J. Moore and B.E. Anderson (ed.) Native warm-season grasses: Research trends and issues. CSSA Spec. Publ. 30. CSSA, Madison, WI.

[19] Lemus R, Brummer E.C, Burras C.L, Moore K.J, Barker M.F, Molstad N.E (2008) Effects of nitrogen fertilization on biomass yield and quality in large fields of established switchgrass in southern Iowa, USA. Biomass and Bioenergy 32: 1187-1194.

[20] Lewandowski I, J.C. Clifton-Brownb J.C, Scurlock J.M.O, Huismand W (2000) Miscanthus: European experience with a novel energy crop. Biomass and Bioenergy 19: 209-227.

[21] Heaton E, Voigt T, Long S.P (2004) A quantitative review comparing the yields of two candidate C4 perennial biomass crops in relation to nitrogen, temperature, and water. Biomass and Bioenergy 27: 21-30.

[22] Heaton E, Dohleman F.G, Long S.P (2008) Meeting US biofuel goals with less land: the potential of Miscnathus. Global Change Bio. 14: 2000-2014.

[23] Madakadze I.C., Coulman B.E, Mcelroy A.R, Stewart K.A, Smith D.L (1998) Evaluation of selected warm-season grasses for biomass production in areas with a short growing season. Bioresource Technology 65: 1-12.

[24] Madakadze I.C., Coulman B.E, Peterson P, Stewart K.A, R. Samson R, Smith D.L (1998) Leaf area development, light interception, and yield among switchgrass populations in a short-season area. Crop Sc. 38: 827-834.

[25] Madakadze I.C, Stewart K, Peterson P.R, Coulman B.E, Smith D.L (1999) Switchgrass biomass and chemical composition for biofuel in eastern Canada. Agron. J. 91: 696-701.

[26] Berdahl J.D., Frank A.B, Krupinsky J.M, Carr P.M, Hanson J.D, Johnson H.A (2005) Biomass yield, phenology, and survival of diverse switchgrass cultivars and experimental strains in western North Dakota. Agron. J. 97: 549-555.

[27] Frank A.B, Berdahl J.D, Hanson J.D, Liebig M.A, Johnson H.A (2004) Biomass and carbon partitioning in switchgrass. Crop Sci. 44: 1391-1396.

[28] Lee D.K, Boe A (2005) Biomass production of switchgrass in central South Dakota. Crop Sc.45: 2583-2590.

[29] Lee D.K, Owens V.N, Doolittle J.J (2007) Switchgrass and soil carbon sequestration response to ammonium nitrate, manure, and harvest frequency on conservation reserve program land. Agron. J. 99: 462-468.

[30] Mulkey V.R, Owens V.N, Lee D.K (2006) Management of switchgrass-dominated conservation reserve program lands for biomass production in South Dakota. Crop Sc. 46: 712-720.

[31] Mulkey V.R, Owens V.N, Lee D.K (2008) Management of warm-season grass mixtures for biomass production in South Dakota USA. Bioresource Tech. 99: 609-617.

[32] Tober D.W, Duckwitz W, Jensen N, Knudson M (2007) Switchgrass biomass trials in North Dakota, South Dakota and Minnesota. USDA-NRCS, Bismarck, ND.

[33] Tober D.W, Jensen N, Duckwitz W, Knudson M (2008) Big bluestem biomass trials in North Dakota, South Dakota and Minnesota. USDA-NRCS, Bismarck, ND.

[34] Kiniry J.R, Schmer M.R, Vogel K.P, Mitchell R.B (2008) Switchgrass biomass simulation at diverse sites in the Northern Great Plains of the U.S. Bioenergy Res. 1: 259-264.

[35] Xue Q, Nyren P.E, Wang G, Eriksmoen E, Bradbury G, Halverson M, Aberle E, Nichols K, Liebig M (2011) Biomass composition of perennial grasses for biofuel production in North Dakota. Biofuels 2: 515-528.

[36] Adler PR, Sanderson MA, Boeteng AA, Weimer PJ, Adler PB, Jung HG (2006) Biomass yield and biofuel quality of switchgrass harvested in fall or spring. Agron. J. 98: 1518–1528.

[37] McKendry P (2002) Energy production from biomass (Part 2): conversion technologies. Bioresource Technology 83: 47–54.

[38] Waramit N, Moore KJ, Haggenstaller AH (2011) Composition of native warm-season grasses for bioenergy production in response to nitrogen fertilization rate and harvest date. Agron. J. 103: 655-662.

[39] Karki B, Nahar N, Pryor S.W (2011).. Enzymatic hydrolysis of switchgrass and tall wheatgrass mixtures using dilute sulfuric acid and aqueous ammonia pretreatments. Biological Engineering 3: 163-171.

[40] Gastine A, J. Roy J, Leadley P.W (2003) Plant biomass production and soil nitrogen in mixtures and monocultures of old field Mediterranean annuals. Acta Oecologia 24: 65-75.

[41] Biondini M (2007) Plant diversity, production, stability, and susceptibility to invasion in restored Northern tall grass prairies (United States). Restoration Ecol. 15: 77-87.

[42] Sedivec K.K, Tober D.W, Berdahl J.D (2001) Grass varieties for North Dakota. NDSU Extension Service. R-794.

[43] Newman P.R, Moser L.E (1988) Grass seedling emergence, morphology, and establishment as affected by planting depth. Agron. J. 80: 383–387.

[44] Nyoka B, Jeranyama P, Boe V, Mooeching M (2007) Management guide for bioass feedstock production from switchgrass in the Northern Great Plains. SGINC2-07. South Dakota State University.

[45] Hsu F.H, Nelson C.J, Matches A.G (1985) Temperature effects on germination of perennial warm-season forage grasses. Crop Sci. 25: 215– 220.

[46] Hanson J.D, Johnson H.A (2005) Germination of switchgrass under different temperature and PH regimes. Seed Tech. J. 27: 203-210.

[47] Parrish D.J, Wolf D.D, Peterson P.R, Daniels W.L (1999) Successful Establishment and Management of Switchgrass. Proceedings of the 2nd Eastern Native Grass Symposium, Baltimore, MD November 1999.

[48] Smart A.J, Moser L.E (1997) Morphological development of switchgrass as affected by planting date. Agron. J. 89: 958–962.

[49] Holmberg K.B, Baldwin S.B (2006) Sequential planting of switchgrass seed to determine optimal planting date for establishment. The ASA Southern Regional Branch Meeting (February 5-7, 2006).

[50] Boe A (2007) Variation between two switchgrass cultivars for components of vegetative and seed biomass. Crop Sci 47:636–642.

[51] Aiken G. E, Springer T.L (1995) Seed size distribution, germination, and emergence of 6 switchgrass cultivars. J. Range Manage. 48: 455–458.

[52] Smart A.J, Moser L.E (1999) Switchgrass seedling development as affected by seed size. Agron. J. 91: 335–338.

[53] Zhang J, Maun M.A (1991) Establishment of *Panicum virgatum* L. seedlings on a Lake Erie sand dune. Bull. Torrey Bot. Club 118:141–153.

[54] Casler   M.D, Undersander D.J (2006) Selection for establishment capacity in reed canarygrass. Crop Sci. 46: 1277-1285.

[55] Richards R.A, Rebetzke G.J, Condon A.G, van Herwaarden A.F (2002) Breeding opportunities for increasing the efficiency of water use and crop yield in temperate cereals. Crop Sci. 42: 111-121.

[56] Hetrick B.A, Kitt D.G, Wilson G.T (1988) Mycorrhizal dependence and growth habit of warm-season and cool- season tallgrass prairie plants. Can. J. Bot. 66: 1376–1380.

[57] Brejda J.J., Moser L.E, Vogel K.P (1998) Evaluation of switchgrass rhizosphere microflora for enhancing seedling yield and nutrient uptake. Agron. J. 90: 753–758.

[58] Hendrickson J.R, Nichols K.A, Johnson H.A (2008) Native and introduced mycorrhizal fungi effect on switchgrass response to water and defoliation stress. IN: Society for Range Management Meeting Abstracts (CD ROM), January 27 - February 1, 2008. Louiville, KY.

[59] Hintz R.L, Harmoney K.R, Moore K.J, George J.R, Brummer E.C (1998) Establishment of switchgrass and big bluestem in corn with atrazine. Agron. J. 90: 591–596.

[60] Cossar R.D, Baldwin B.S (2004) Establishment of switchgrass with orghum-sudangrass. In: Randall J, Burns J.C., editors, Proc. Third Eastern Native Grass Symposium Omnipress, Chapel Hill, NC. pp. 98-102.

[61] Mitchell R.B, Britton C.M (2000) Managing weeds to establish and maintain warm-season grasses. In: *Native Warm-Season Grasses: Research Trends and Issues.*, pp. 159–176.

Anderson, B. E. and Moore, K. J., Eds., CSSA Special Pub. No. 30. Crop Science Society of America, Madison, WI.

[62] Bovey R.W., Hussey M.A (1991) Response of selected forage grasses to herbicides. Agron. J. 83: 709–713.

[63] Buhler D.D, Netzer D.A, Riemenschneider D.E, Hartzler R.G (1998) Weed management in short rotation poplar and herbaceous perennial crops grown for biofuel production. Biomass & Bioenergy 14: 385-394.

[64] O'Donovan J.T, Newman J.C, Harker K.N, Blackshaw R.E, and D. W. McAndrew D.W (1999) Effect of barley plant density on wild oat interference, shoot biomass and seed yield under zero tillage. Can. J. Plant Sci 79:655–662.

[65] O'Donovan J.T, Harker K.N, Clayton G.W, Hall L.M (2000) Wild oat (*Avena fatua*) interference in barley (Hordeum vulgare) is influenced by barley variety and seeding rate. Weed Technol 14:624–629.

[66] O'Donovan J. T, Blackshaw R.E, Harker K.N, Clayton G.W (2006) Wheat Seeding Rate Influences Herbicide Performance in Wild Oat (Avena fatua L.). Agron. J. 98:815-822.

[67] Xue Q, Stougaard R.N (2002) Spring wheat seed size and seeding rate affect wild oat demographics. Weed Sci 50:312–320.

[68] Xue Q, Stougaard R.N (2006) Effects of spring wheat seed size and reduced rates of tralkoxydim on wild oat control, wheat yield, and economic returns. Weed Tech. 20: 472-477.

[69] Vogel K.P, Brejda J.J, Walters D.T, Buxton D.R (2002) Switchgrass biomass production in the Midwest USA: harvest and nitrogen management. Agron. J. 94:413–420.

[70] Muir J.P, Sanderson M.A, Ocumpaugh W.R, Jones R.M, Reed R.L (2001) Biomass production of 'Alamo' switchgrass in response to nitrogen, phosphorus, and row spacing. Agron. J. 93:896–901.

[71] Gunderson C.A, Davis E.B, Jager H.I, West T.O, Perlack R.D, Brandt C.C, Wullschleger S.D, Baskaran L.M, Wilkerson E.G, Downing M.E (2008) Exploring Potential U.S. Switchgrass Production for Lignocellulosic Ethanol. Oakridge National Laboratory Pub. ORNL/TM-2007/183.

[72] Clark F.E (1977) Internal cycling of 15N in shortgrass prairie. Ecology 58:1322–1333.

[73] Sanderson M.A, Read J.C, Reed R.L (1999) Harvest management of switchgrass for siomass seedstock and forage production. Agron. J. 91: 5-10.

[74] Guretzky J.A, Biermacher J.T, Cook B.J, Kering M.K, Mosali J (2011) Switchgrass for forage and bioenergy: harvest and nitrogen rate effects on biomass yields and nutrient composition. Plant Soil 339: 69-81.

[75] Nyren P.E, Wang G, Patton B, Xue Q, Bradbury G, Halvorson M, Aberle E (2012) Evaluation of Perennial Forages for Use as Biofuel Crops in Central and Western North Dakota.http://www.ag.ndsu.edu/CentralGrasslandsREC/biofuels-research-1/2011-report/Biomass_for_ethanol.pdf (accessed on April 15, 2012).

[76] Boehmel C, Lewandowski I, Claupein W (2008) Comparing annual and perennial energy cropping systems with different management intensities. Agric. Systems 96: 224-236.

[77] Anex R.P, Lynd L.R, Laser M.S, Heggenstaller A.H, Liebman M (2007) Potential for enhanced nutrient cycling through coupling of agricultural and bioenergy systems. Crop Sci. 47:1327-1335.

[78] Roth A.M, Sample D.W, Ribic C.A, Paine, Undersander D.J, Bartelt G.A (2005). Grassland bird response to harvesting switchgrass as a biomass energy crop. Biomass and Bioenergy 28: 490-498.

Chapter 3

Design of a Cascade Observer for a Model of Bacterial Batch Culture with Nutrient Recycling

Miled El Hajji and Alain Rapaport

Additional information is available at the end of the chapter

1. Introduction

Microbial growths and their use for environmental purposes, such as bio-degradations, are widely studied in the industry and research centres. Several models of microbial growth and bio-degradation kinetic have been proposed and analysed in the literature. The Monod's model is one of the most popular ones that describes the dynamics of the growth of a biomass of concentration X on a single substrate of concentration S in batch culture [15, 18]:

$$\dot{S} = -\frac{\mu(S)}{Y} X, \quad \dot{X} = \mu(S) X.$$ (1)

where the specific growth rate $\mu(\cdot)$ is:

$$\mu(S) = \mu_{max} \frac{S}{K_s + S},$$ (2)

with μ_{max}, K_s and Y are repsectively the maximum specific growth rate, the affinity constant and the yield coefficient. Other models take explicitly into account a lag-phase before the growth, such as the Baranyi's [1–3] or the Buchanam's [6] ones. These models are well suited for the growth phase (i.e. as long as a substantial amount of substrate remains to be converted) but not after [18], because the accumulation of dead or non-viable cells is not taken into account. Part of the non-viable cells release substrate molecules, in quantities that are no longer negligible when most of the initial supply has been consumed. The on-line observation of the optical density of the biomass provides measurements of the total biomass, but not of the proportion among dead and viable cells. Some tools allow the distinction between viable and dead cells but do not detect non-viable non-dead ones [22].

In this work, we consider an extension of the model (1) considering both the accumulation of dead cells and the recycling of part of it into substrate, and tackle the question of parameters and state reconstruction. To our knowledge, this kind of question has not been thoroughly studied in the literature. Models of continuous culture with nutrient recycling have already been studied [4, 5, 9, 12–14, 16, 20, 21, 24, 25] but surprisingly few works considers batch cultures. A possible explanation comes from the fact that only the first stage of the growth, for which cell mortality and nutrient recycling can be neglected, is interested for industrial applications. Nevertheless, in natural environment such as in soils, modelling the growth end is also important, especially for biological decontamination and soil bioremediation.

Moreover, we face a model for which the parameters are not identifiable at steady state. Then, one cannot apply straightforwardly the classical estimation techniques, that usually requires the global observability of the system. Estimation of parameters in growth models, such as the Baranyi's one, are already known to be difficult to tackle in their differential form [11]. In addition, we aim here at reconstructing on-line unmeasured state variables (amounts of viable and non-viable cells), as well as parameters. For this purpose, we propose the coupling of two non-linear observers in cascade with different time scales, providing a practical convergence of the estimation error. Design of cascade observers in biotechnology can be found for instance in [17, 23], but with the same time scale.

2. Derivation of the model

We first consider a mortality rate in the model (1):

$$\dot{X} = \mu(S)X - mX$$

where parameter $m > 0$ becomes not negligible when $\mu(S)$ takes small values. In addition, we consider an additional compartment X_d that represents the accumulation of dead cells:

$$\dot{X}_d = \delta m X,$$

where the parameter $\delta \in (0, 1)$ describes the part of non-viable cells that are not burst. We assume that the burst cells recycle part of the substrate that has been assimilated but not yet transformed. Then, the dynamics of the substrate concentration can be modified as follows:

$$\dot{S} = -\frac{\mu(S)}{Y}X + \lambda(1 - \delta)mX,$$

where $\lambda > 0$ is recycling conversion factor. It appears reasonable to assume that the factor λ is smaller that the growth one:

Assumption A1. $\dfrac{1}{Y} > \lambda$.

In the following we assume that the growth function $\mu(\cdot)$ and the yield coefficient Y of the classical Monod's model are already known. Typically, they can be identified by measuring the initial growth slope on a series of experiments with viable biomass and different initial concentrations, mortality being considered to be negligible during the exponential growth. We aim at identifying the three parameters m, δ and λ, and on-line reconstructing the variables

X and X_d, based on on-line observations of the substrate concentration S and the total biomass $B = X + X_d$.

Without any loss of generality, we shall assume that the growth function $\mu(\cdot)$ can be any function satisfying the following hypotheses.

Assumption A2. The function $\mu(\cdot)$ is a smooth increasing function with $\mu(0) = 0$.

For sake of simplicity, we normalise several quantities, defining

$$s = S, \; x = X/Y, \; x_d = X_d/Y, \; a = (1-\delta)m \text{ and } k = \lambda Y.$$

Then, our model can be simply written as

$$\begin{cases} \dot{s} = -\mu(s)x + kax, \\ \dot{x} = \mu(s)x - mx, \\ \dot{x}_d = mx - ax, \end{cases} \tag{3}$$

along with the observation vector $y = \begin{pmatrix} s \\ x + x_d \end{pmatrix}$. Typically, we consider known initial conditions such that

$$s(0) = s_0 > 0, \quad x_d(0) = 0 \text{ and } x(0) = x_0 > 0.$$

Our purpose is to reconstruct parameters m, a and k and state variable $x(\cdot)$ or $x_d(\cdot)$, under the constraints $m > a$ and $k < 1$, that are direct consequences of the definition of a and Assumption A1. Moreover, we shall assume that a priori bounds on the parameters are known i.e.

$$(m, a, k) \in [m^-, m^+] \times [a^-, a^+] \times [k^-, k^+]. \tag{4}$$

3. Properties of the model

Proposition 1. *The dynamics (3) leaves invariant the 3D-space $\mathcal{D} = \mathbb{R}^3_+$ and the set*

$$\Omega = \left\{ (s, x, x_d) \in \mathcal{D} \mid s + x + \frac{(m-ka)}{(m-a)} x_d = s_0 + x_0 \right\}.$$

Proof. The invariance of $\mathbb{R}^3_+$ is guaranteed by the following properties:

$$s = 0 \;\Rightarrow\; \dot{s} = kax \geq 0,$$
$$x = 0 \;\Rightarrow\; \dot{x} = 0,$$
$$x_d = 0 \;\Rightarrow\; \dot{x}_d = (m-a)x \geq 0.$$

Consider the quantity $M = s + x + \frac{(m-ka)}{(m-a)} x_d$. One can easily check from equations (3) that one has $\dot{M} = 0$, leading to the invariance of the set Ω. ∎

Let $\bar{s}$ be the number $\bar{s} = \mu^{-1}(m)$ or $+\infty$.

Proposition 2. *The trajectories of dynamics (3) converge asymptotically toward an equilibrium point*

$$E^\star = \left(s^\star, 0, \frac{m-a}{m-ka}(s_0 + x_0 - s^\star) \right)$$

with $s^\star \le \min(s_0 + x_0, \bar{s})$.

Proof. The invariance of the set Ω given in Proposition 1 shows that all the state variables remain bounded. From equation $\dot{x}_d = (m-a)x$ with $m > a$, and the fact that x_d is bounded, one deduces that $x(\cdot)$ has to converge toward 0, and $x_d(\cdot)$ is non increasing and converges toward $x_d^\star$ such that $x_d^\star \in [0, (s_0 + x_0)(m-a)/(m-ka)]$. Then, from the invariant defined by the set Ω, $s(\cdot)$ has also to converges to some $s^\star \le s_0 + x_0$. If $s^\star$ is such that $s^\star > \bar{s}$, then from equation $\dot{x} = (\mu(s) - m)x$, one immediately see that $x(\cdot)$ cannot converge toward 0. ∎

4. Observability of the model

We recall that our aim is to estimate on-line both parameters and unmeasured variables x, x_d, based on the measurements. One can immediately see from equations (3) that parameters (m, a, k) cannot be reconstructed observing the system at steady state. Nevertheless, considering the derivative μ' of μ with respect to s and deriving the outputs:

$$\begin{cases} \dot{y}_1 = (-\mu(y_1) + ka)\,x, \\ \dot{y}_2 = (\mu(y_1) - a)\,x, \\ \ddot{y}_1 = (\mu(y_1) - m)\,\dot{y}_1 - \mu'(y_1)\,x\,\dot{y}_1, \\ \ddot{y}_2 = (\mu(y_1) - m)\,\dot{y}_2 + \mu'(y_1)\,x\,\dot{y}_1., \end{cases}$$

one obtains explicit expression of the parameters and unmeasured state variable as functions of the outputs and its derivatives, away from steady state:

$$\begin{cases} m = \mu(y_1) - \dfrac{\ddot{y}_1 + \ddot{y}_2}{\dot{y}_1 + \dot{y}_2}, \\[2mm] x = \dfrac{\ddot{y}_2 - (\mu(y_1) - m)\dot{y}_2}{\mu'(y_1)\dot{y}_1}, \\[2mm] x_d = y_2 - x, \\[2mm] a = \mu(y_1) - \dfrac{\dot{y}_2}{x}, \\[2mm] k = \dfrac{\mu(y_1)}{a} + \dfrac{\dot{y}_1}{a\,x}, \end{cases} \tag{5}$$

from which one deduces the observability of the system.

5. Design of a practical observer

Playing with the structure of the dynamics, we are able to write the model as a particular cascade of two sub-models. We first present a practical observer for the reconstruction of the parameters a and k using the observation y_1 only, but with a change of time that depends on y_1 and y_2. We then present a second observer for the reconstruction of the parameter m

and the state variables x and x_d, using both observations y_1 and y_2 and the knowledge of the parameters a and k. Finally, we consider the coupling of the two observers, the second one using the estimations of a and k provided by the first one. More precisely, our model is of the form

$$\dot{Z} = F(Z,P) \quad , \quad y = H(Z)$$

where F is our vector field with the state, parameters and observation vectors Z, P and y of dimension respectively 3, 3 and 2. We found a partition

$$Z = \begin{pmatrix} Z_1 \\ Z_2 \end{pmatrix}, P = \begin{pmatrix} P_1 \\ P_2 \end{pmatrix} \text{ s.t. } \begin{cases} \dim Z_1 = 1, \dim P_1 = 2 \\ \dim Z_2 = 2, \dim P_2 = 1 \end{cases}$$

$$y = \begin{pmatrix} y_1 \\ y_2 \end{pmatrix} = \begin{pmatrix} H_1(Z_1) \\ H_2(Z_2) \end{pmatrix}$$

and the dynamics is decoupled as follows

$$\dot{Z}_1 = \frac{1}{\frac{d\phi(y)}{dt}} F_1(Z_1, P_1)$$

$$\dot{Z}_2 = F_2(Z_2, y_1, P_1, P_2)$$

with $d\phi(y)/dt > 0$. Moreover, the following characteristics are fulfilled:

i. (Z_1, P_1) is observable for the dynamics (F_1, H_1) i.e. without the term $d\phi(y)/dt$,

ii. (Z_2, P_2) is observable for the dynamics (F_2, H_2) when P_1 is known. Then, the consideration of two observers $\hat{F}_1(\cdot)$ and $\hat{F}_2(P_1, \cdot)$ for the pairs (Z_1, P_1) and (Z_2, P_2) respectively, leads to the construction of a cascade observer

$$\frac{d}{d\tau} \begin{pmatrix} \hat{Z}_1 \\ \hat{P}_1 \end{pmatrix} = \hat{F}_1(\hat{Z}_1, \hat{P}_1, y_1),$$

$$\frac{d}{dt} \begin{pmatrix} \hat{Z}_2 \\ \hat{P}_2 \end{pmatrix} = \hat{F}_2(\hat{P}_1, \hat{Z}_2, \hat{P}_2, y_2)$$

with $\tau(t) = \phi(y(t)) - \phi(y(0))$, that we make explicit below. Notice that the coupling of two observers is made by $\hat{P}_1$, and that the term $d\phi(y)/dt$ prevents to have an asymptotic convergence when $\lim_{t \to +\infty} \tau(t) < +\infty$.

Definition 1. *An estimator $\hat{Z}_\gamma(\cdot)$ of a vector $Z(\cdot)$, where $\gamma \in \Gamma$ is a parameter, is said to have a practical exponential convergence if there exists positive constants K_1, K_2 such that for any $\epsilon > 0$ and $\theta > 0$, the inequality*

$$\|\hat{Z}_\gamma(t) - Z(t)\| \le \epsilon + K_1 e^{-K_2 \theta t}, \quad \forall t \ge 0$$

is fulfilled for some $\gamma \in \Gamma$.

In the following we shall denote by $\text{sat}(l, u, \iota)$ the saturation operator $\max(l, \min(u, \iota))$.

5.1. A first practical observer for k and a

Let us consider the new variable

$$\tau(t) = y_1(0) - y_1(t) + y_2(0) - y_2(t)$$

that is measured on-line. From Proposition 1, one deduces that $\tau(\cdot)$ is bounded. One can also easily check the property

$$\frac{d\tau}{dt} = (1-k)\,a\,x(t) > 0, \quad \forall t \geq 0.$$

Consequently, $\tau(\cdot)$ is an increasing function up to

$$\bar{\tau} = \lim_{t \to +\infty} \tau(t) < +\infty \tag{6}$$

and $\tau(\cdot)$ defines a diffeomorphism from $[0, +\infty)$ to $[0, \bar{\tau})$. Then, one can check that the dynamics of the variable s in time τ is decoupled from the dynamics of the other state variables:

$$\frac{ds}{d\tau} = \alpha - \beta\mu(s)$$

where α and β are parameters defined as combinations of the unknown parameters a and k:

$$\alpha = \frac{k}{1-k},$$

$$\beta = \frac{1}{a(1-k)}$$

and from (4) one has $(\alpha, \beta) \in [\alpha^-, \alpha^+] \times [\beta^-, \beta^+]$. For the identification of the parameters α, β, we propose below to build an observer. Other techniques, such as least squares methods, could have been chosen. An observer presents the advantage of exhibiting a innovation vector that gives a real-time information on the convergence of the estimation.

Considering the state vector $\xi = \left[s \;\; \dfrac{ds}{d\tau} \;\; \dfrac{d^2s}{d\tau^2} \right]^T$, one obtains the dynamics

$$\frac{d\xi}{d\tau} = A\xi + \begin{pmatrix} 0 \\ 0 \\ \varphi(y_1, \xi) \end{pmatrix} \quad \text{with } y_1 = C\xi,$$

$$\varphi(y_1, \xi) = \frac{\xi_3^2}{\xi_2} + \xi_2\xi_3\frac{\mu''(y_1)}{\mu'(y_1)},$$

and the pair (A, C) in the Brunovsky's canonical form:

$$A = \begin{pmatrix} 0\,1\,0 \\ 0\,0\,1 \\ 0\,0\,0 \end{pmatrix} \quad \text{and } C = \begin{pmatrix} 1\,0\,0 \end{pmatrix}. \tag{7}$$

The unknown parameters α and β can then be made explicit as functions of the observation y_1 and the state vector ξ:

$$\alpha = l_\alpha(y_1, \xi) = \xi_2 - \frac{\xi_3 \mu(y_1)}{\xi_2 \mu'(y_1)},$$

$$\beta = l_\beta(y_1, \xi) = -\frac{\xi_3}{\xi_2 \mu'(y_1)}.$$

One can notice that functions $\varphi(y_1, \cdot)$, $l_\alpha(y_1, \cdot)$ and $l_\beta(y_1, \cdot)$ are not well defined on $\mathbb{R}^3$, but along the trajectories of (3) one has $\xi_3/\xi_2 = -\beta\mu'(y_1)$ and $\xi_2 = \alpha - \beta\mu(y_1)$, that are bounded. Moreover Assumption A2 guarantees that $\mu'(y_1)$ is always strictly positive . We can consider (globally) Lipschitz extensions of these functions away from the trajectories of the system, as follows:

$$\tilde{\varphi}(y_1, \xi) = \xi_3 \left(h_1(y_1, \xi) + \frac{\mu''(y_1)}{\mu'(y_1)} h_2(y_1, \xi) \right),$$

$$\tilde{l}_\alpha(y_1, \xi) = \xi_2 - h_1(y_1, \xi) \frac{\mu(y_1)}{\mu'(y_1)},$$

$$\tilde{l}_\beta(y_1, \xi) = -\frac{h_1(y_1, \xi)}{\mu'(y_1)}$$

with

$$h_1(y_1, \xi) = \text{sat}\left(-\beta^+\mu'(y_1), -\beta^-\mu'(y_1), \frac{\xi_3}{\xi_2} \right),$$

$$h_2(y_1, \xi) = \text{sat}\left(\alpha^- - \beta^+\mu(y_1), \alpha^+ - \beta^-\mu(y_1), \xi_2 \right).$$

Then one obtains a construction of a practical observer.

Proposition 3. *There exist numbers $b_1 > 0$ and $c_1 > 0$ such that the observer*

$$\frac{d\hat{\xi}}{d\tau} = A\hat{\xi} + \begin{pmatrix} 0 \\ 0 \\ \tilde{\varphi}(y_1, \hat{\xi}) \end{pmatrix} - \begin{pmatrix} 3\theta_1 \\ 3\theta_1^2 \\ \theta_1^3 \end{pmatrix} (\hat{\xi}_1 - y_1) \tag{8}$$

$$(\hat{\alpha}, \hat{\beta}) = \left(\tilde{l}_\alpha(y_1, \hat{\xi}), \tilde{l}_\beta(y_1, \hat{\xi}) \right)$$

guarantees the convergence

$$\max \left(|\hat{\alpha}(\tau) - \alpha|, |\hat{\beta}(\tau) - \beta| \right) \leq b_1 e^{-c_1 \theta_1 \tau} \|\hat{\xi}(0) - \xi(0)\| \tag{9}$$

for any θ_1 large enough and $\tau \in [0, \bar{\tau})$.

Proof. Consider a trajectory of dynamics (3) and let $O_1 = \{y_1(t)\}_{t\geq0}$. From Proposition 1, one knows that the set O_1 is bounded.

Define $K_{\theta_1} = -\left(3\theta_1 \; 3\theta_1^2 \; \theta_1^3 \right)^T$. One can check that $K_{\theta_1} = -P_{\theta_1}^{-1} C^T$, where P_{θ_1} is solution of the algebraic equation

$$\theta_1 P_{\theta_1} + A^T P_{\theta_1} + P_{\theta_1} A = C^T C.$$

Consider then the error vector $e = \hat{\zeta} - \zeta$. One has

$$\frac{de}{d\tau} = (A + K_{\theta_1} C)e + \begin{pmatrix} 0 \\ 0 \\ \tilde{\varphi}(y_1, \hat{\zeta}) - \tilde{\varphi}(y_1, \zeta) \end{pmatrix}$$

where $\tilde{\varphi}(y_1, \cdot)$ is (globally) Lipschitz on $\mathbb{R}^3$ uniformly in $y_1 \in O_1$. We then use the result in [10] that provides the existence of numbers $c_1 > 0$ and $q_1 > 0$ such that $||e(\tau)|| \leq q_1 e^{-c_1 \theta_1 \tau} ||e(0)||$ for θ_1 large enough. Finally, functions $\tilde{l}_\alpha(y_1, \cdot)$, $\tilde{l}_\beta(y_1, \cdot)$ being also (globally) Lipschitz on $\mathbb{R}^3$ uniformly in $y_1 \in O_1$, one obtains the inequality (9). ∎

Corollary 1. *Estimation of a and k with the same convergence properties than (9) are given by*

$$\hat{k}(\tau), \hat{a}(\tau) = sat\left(k^-, k^+, \frac{\hat{\alpha}(\tau)}{1 + \hat{\alpha}(\tau)}\right), sat\left(a^-, a^+, \frac{1 + \hat{\alpha}(\tau)}{\hat{\beta}(\tau)}\right)$$

Remark. The observer (8) provides only a practical convergence since $\tau(t)$ does not tend toward $+\infty$ when the time t get arbitrary large. For large values of initial x, it may happens that $\mu(t) > t$ for some times $t > 0$. Because the present observer requires the observation y_1 until time τ, it has to be integrated up to time $\min(\tau(t), t)$ when the current time is t.

5.2. A second observer for m and x

We come back in time t and consider the measured variable $z = y_1 + y_2$. When the parameters α and β are known, the dynamics of the vector $\zeta = \begin{bmatrix} z & \dot{z} & \ddot{z} \end{bmatrix}^T$ can be written as follows:

$$\dot{\zeta} = A\zeta + \begin{pmatrix} 0 \\ 0 \\ \psi(y_1, \zeta, \alpha, \beta) \end{pmatrix} \text{ with } z = C\zeta$$

and $\psi(y_1, \zeta, \alpha, \beta) = \dfrac{\zeta_3^2}{\zeta_2} + \zeta_2^2 \mu'(y_1)(\beta\mu(y_1) - \alpha)$

Parameter m and variable $x(\cdot)$ can then be made explicit as functions of y_1 and ζ:

$$m = l_m(y_1, \zeta) = \mu(y_1) - \frac{\zeta_3}{\zeta_2}, \quad x = -\beta\zeta_2$$

Functions $\psi(y_1, \cdot, \alpha, \beta)$ and $l_m(y_1, \cdot)$ are not well defined in $\mathbb{R}^3$ but along the trajectories of the dynamics (3), one has $\zeta_3/\zeta_2 = \mu(y_1) - m$ and $\zeta_2 = -x/\beta$ that are bounded. These functions can be extended as (globally) Lipschitz functions w.r.t. ζ:

$$\tilde{\psi}(y_1, \zeta, \alpha, \beta) = h_3(y_1, \zeta)\zeta_3 + \min(\zeta_2^2, z(0)^2/\beta^2)\mu'(y_1)(\beta\mu(y_1) - \alpha)$$

$$(10)$$

$$\tilde{l}_m(y_1, \zeta) = \mu(y_1) - h_3(y_1, \zeta)$$

with

$$h_3(y_1, \zeta) = \text{sat}\left(\mu(y_1) - m^+, \mu(y_1) - m^-, \frac{\zeta_3}{\zeta_2}\right).$$

Proposition 4. *When α and β are known, there exists numbers $b_2 > 0$ and $c_2 > 0$ such that the observer*

$$\frac{d}{dt}\hat{\zeta} = A\hat{\zeta} + \begin{pmatrix} 0 \\ 0 \\ \tilde{\psi}(y_1, \hat{\zeta}, \alpha, \beta) \end{pmatrix} - \begin{pmatrix} 3\theta_2 \\ 3\theta_2^2 \\ \theta_2^3 \end{pmatrix}(\hat{\zeta}_1 - y_1 - y_2) \tag{11}$$

$$(\hat{m}, \hat{x}) = (\tilde{I}_m(y_1, \hat{\zeta}), -\beta\hat{\zeta}_2)$$

guarantees the exponential convergence

$$\max\left(|\hat{m}(t) - m|, |\hat{x}(t) - x(t)|\right) \leq b_2 e^{-c_2\theta_2 t}||\hat{\zeta}_2(0) - \zeta_2(0)||$$

for any θ_2 large enough and $t \geq 0$.

Proof. As for the proof of Proposition 3, it is a straightforward application of the result given in [10]. ∎

5.3. Coupling the two observers

We consider now the coupling of observer (11) with the estimation $(\hat{\alpha}, \hat{\beta})$ provided by observer (8). This amounts to study the robustness of the second observer with respect to uncertainties of parameters α and β.

Proposition 5. *Consider the observer (11) with (α, β) replaced by $(\tilde{\alpha}(\cdot), \tilde{\beta}(\cdot))$ such that*

$$(\tilde{\alpha}(t), \tilde{\beta}(t)) \in [\alpha^-, \alpha^+] \times [\beta^-, \beta^+], \quad \forall t \geq 0,$$

then there exists positive numbers $\bar{b}_2, \bar{c}_2, \bar{d}_2$ such that for any $\epsilon > 0$ there exists θ_2 large enough to guarantee the inequalities

$$|\hat{m}(t) - m| \leq \epsilon + \bar{b}_2 e^{-\bar{c}_2 t}||\hat{\zeta}(0) - \zeta(0)|| \tag{12}$$

$$|\hat{x}(t) - x(t)| \leq \epsilon + \bar{d}_2|\tilde{\beta}(t) - \beta| + \bar{b}_2 e^{-\bar{c}_2 t}||\hat{\zeta}(0) - \zeta(0)|| \tag{13}$$

for any $t \geq 0$.

Proof. As for the proof of Proposition 3, we fix an initial condition of system (3) and consider the bounded set $O_1 = \{y_1(t)\}_{t \geq 0}$. The dynamics of $e = \hat{\zeta} - \zeta$ is

$$\dot{e} = (A + K_{\theta_2}C)e + (\tilde{\psi}(y_1, \hat{\zeta}, \tilde{\alpha}, \tilde{\beta}) - \tilde{\psi}(y_1, \zeta, \alpha, \beta))v$$

where (A, C) in the Brunovsky's form (7), $v = \begin{pmatrix} 0 & 0 & 1 \end{pmatrix}^T$ and $K_{\theta_2} = -P_{\theta_2}^{-1} C^T$ with

$$P_{\theta_2} = \begin{pmatrix} \theta_2^{-1} & -\theta_2^{-2} & \theta_2^{-3} \\ -\theta_2^{-2} & 2\theta_2^{-3} & -3\theta_2^{-4} \\ \theta_2^{-3} & -3\theta_2^{-4} & 6\theta_2^{-5} \end{pmatrix} \tag{14}$$

solution of the algebraic equation

$$\theta_2 P_{\theta_2} + A^T P_{\theta_2} + P_{\theta_2} A = C^T C. \tag{15}$$

Consider then $V(t) = ||e(t)||_{P_{\theta_2}}^2 = e^T(t) P_{\theta_2} e(t)$. Using (15), one has

$$\begin{aligned} \dot{V} &= -\theta_2 e^T P_{\theta_2} e - e^T C^T C e + 2\delta e^T P_{\theta_2} v \\ &\leq -\theta_2 ||e||_{P_{\theta_2}}^2 + 2\delta ||e||_{P_{\theta_2}} ||v||_{P_{\theta_2}} \end{aligned} \tag{16}$$

where $\delta = |\tilde{\psi}(y_1, \hat{\zeta}, \tilde{\alpha}, \tilde{\beta}) - \tilde{\psi}(y_1, \zeta, \alpha, \beta)|$.

One can easily compute from (14) $||v||_{P_{\theta_2}} = \sqrt{6}\theta^{-5/2}$.

From the expression (10) and the (globally) Lipschitz property of the map $\zeta \mapsto \tilde{\psi}(y_1, \zeta, \alpha, \beta)$ uniformly in $y_1 \in O_1$, we deduce the existence of two positive numbers c and L such that

$$\begin{aligned} \delta &\leq |\tilde{\psi}(y_1, \hat{\zeta}, \tilde{\alpha}, \tilde{\beta}) - \tilde{\psi}(y_1, \hat{\zeta}, \alpha, \beta)| + |\tilde{\psi}(y_1, \hat{\zeta}, \alpha, \beta) - \tilde{\psi}(y_1, \zeta, \alpha, \beta)| \\ &\leq |\tilde{\psi}(y_1, \hat{\zeta}, \alpha^-, \beta^+) - \tilde{\psi}(y_1, \hat{\zeta}, \alpha^+, \beta^-)| + L||e|| \\ &\leq c + L||e|| \end{aligned} \tag{17}$$

Notice that one has $||e||_{P_{\theta_2}} = \theta_2||\tilde{e}||_{P_1}$ with $\tilde{e}_i = \theta_2^{-i} e_i$ and $||\tilde{e}||^2 \geq \theta_2^{-6}||e||^2$ for any $\theta_2 \geq 1$. The norms $|| \cdot ||_{P_1}$ and $|| \cdot ||$ being equivalent, there exists a numbers $\eta > 0$ such that $||\tilde{e}||_{P_1}|| \geq \eta||\tilde{e}||$, and we deduce the inequality

$$||e||_{P_{\theta_2}} \geq \eta\theta_2^{-5/2}||e||. \tag{18}$$

Finally, gathering (16), (17) and (18), one can write

$$\frac{d}{dt}||e||_{P_{\theta_2}} \leq \left(-\frac{\theta_2}{2} + \frac{\sqrt{6}L}{\eta} \right) ||e||_{P_{\theta_2}} + \sqrt{6}\theta_2^{-5/2} c$$

For θ_2 large enough, one has $-\theta_2/2 + \sqrt{6}L/\eta < 0$ and then, using again (18), obtains

$$\frac{d}{dt}||e|| \leq \left(-\frac{\theta_2}{2} + \frac{\sqrt{6}L}{\eta} \right) ||e|| + \frac{\sqrt{6}}{\eta} c$$

from which we deduce the exponential convergence of the error vector e toward any arbitrary small neighbourhood of 0 provided that θ_2 is large enough.

The Lipschitz continuity of the map $l_m(\cdot)$ w.r.t. ζ uniformly in $y_1 \in O_1$ provides the inequality (12).

For the estimation of $x(\cdot)$, one has the inequality

$$|\hat{x} - x| = |\hat{\beta}\hat{\zeta}_2 - \beta\zeta_2| \leq |\hat{\beta} - \beta||\zeta_2| + \beta^+|\hat{\zeta}_2 - \zeta_2|$$

provided the estimation (13), the variable ζ_2 being bounded. ∎

Corollary 2. *At any time $t > 0$, the coupled observer*

$$\frac{d\hat{\zeta}}{ds_1} = A\hat{\zeta} + \begin{pmatrix} 0 \\ 0 \\ \tilde{\varphi}(y_1,\hat{\zeta}) \end{pmatrix} - \begin{pmatrix} 3\theta_1 \\ 3\theta_1^2 \\ \theta_1^3 \end{pmatrix}(\hat{\zeta}_1 - y_1)$$

$$\frac{d\hat{\zeta}}{ds_2} = A\hat{\zeta} + \begin{pmatrix} 0 \\ 0 \\ \tilde{\psi}(y_1,\hat{\zeta},\hat{\alpha}(s_2),\hat{\beta}(s_2)) \end{pmatrix} - \begin{pmatrix} 3\theta_2 \\ 3\theta_2^2 \\ \theta_2^3 \end{pmatrix}(\hat{\zeta}_1 - y_1 - y_2)$$

integrated for $s_1 \in [0, \min(t, \tau(t))]$ and $s_2 \in [0, t]$, with

$$\tau(t) = y_1(0) - y_1(t) + y_2(0) - y_2(t),$$
$$\hat{\alpha}(s_2) = sat(\alpha^-, \alpha^+, \tilde{l}_\alpha(y_1(\min(s_2, \tau(t)))), \hat{\zeta}(\min(s_2, \tau(t)))),$$
$$\hat{\beta}(s_2) = sat(\beta^-, \beta^+, \tilde{l}_\beta(y_1(\min(s_2, \tau(t)))), \hat{\zeta}(\min(s_2, \tau(t)))),$$

provides the estimations

$$\hat{m}(t) = \tilde{l}_m(y_1(t), \hat{\zeta}(t)),$$
$$(\hat{x}(t), \hat{x}_d(t)) = (-\hat{\beta}(t)\hat{\zeta}_2(t), y_2(t) + \hat{\beta}(t)\hat{\zeta}_2(t)).$$

The convergence of the estimator is exponentially practical, provided θ_1 and θ_2 to be sufficiently large.

6. Numerical simulations

We have considered a Monod's growth function (2) with the parameters $\mu_{max} = 1$ and $K_s = 100$ and the initial conditions $s(0) = 50$, $x(0) = 1$, $x_d(0) = 0$. The parameters to be reconstructed have been chosen, along with a priory bounds, as follows:

parameter	δ	k	m
value	0.2	0.2	0.1
bounds	$[0.1, 0.3]$	$[0.1, 0.3]$	$[0.05, 0.2]$

Those values provide an effective growth that is reasonably fast ($s(0)$ is about $K_s/2$), and a value $\bar{\tau}$ (see (6)) we find by numerical simulations is not too small. For the time interval $0 \leq t \leq t_{max} = 80$, we found numerically the interval $0 \leq \tau \leq \tau_{max} = \tau(tmax) \simeq 37.22$ (see Figure 1). For the first observer, we have chosen a gain parameter $\theta_1 = 3$ that provides

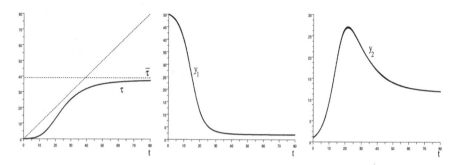

Figure 1. Graphs of function τ and observations y_1, y_2.

a small error on the estimation of the parameters α and β at time τ_{max} (see Figures 2 and 3). These estimations have been used on-line by the second observer, with $\theta_2 = 2$ as a choice

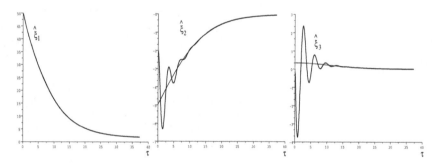

Figure 2. Internal variables $\hat{\zeta}$ of the first observer in time τ (variables ζ of the true system in thin lines).

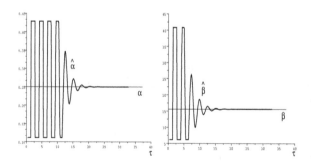

Figure 3. On-line estimation of parameters α and β.

for the gain parameter. On Figures 4 and 5, one can see that the estimation error get small when the estimations provided by the first observer are already small.   Simulations have been also conducted with additive noise on measurements y_1 and y_2 with a signal-to-noise

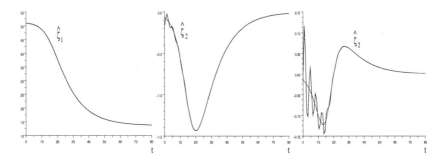

Figure 4. Internal variables $\hat{\zeta}$ of the second observer in time t (variables ζ of the true system in thin lines).

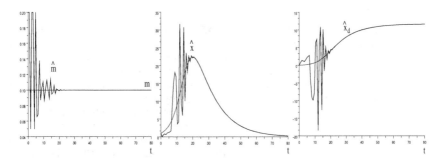

Figure 5. On-line estimation of parameter m and state variables x and x_d.

ratio of 10 and a frequency of $0.1Hz$ (see Figures 6 and 7).   In presence of a low frequency

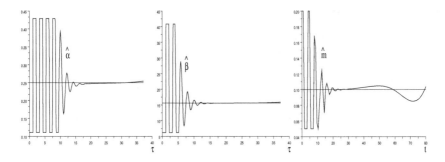

Figure 6. Estimation of the parameters α, β and m in presence of measurement noise.

noise (as it can be usually assumed in biological applications), one finds a good robustness of the estimations of parameters α, β and variables x and x_d. Estimation of parameter m is more affected by noise. This can be explained by the structure of the equations (5): the estimation of m is related to the second derivative of both observations y_1 and y_2, and consequently is more sensitive to noise on the observations.

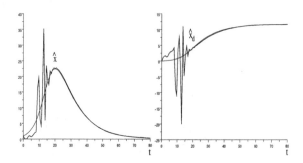

Figure 7. Estimation of the state variables x and x_d in presence of measurement noise.

7. Conclusion

The extension of the Monod's model with an additional compartment of dead cells and substrate recycling terms is no longer identifiable, considering the observations of the substrate concentration and the total biomass. Nevertheless, we have shown that the model can be written in a particular cascade form, considering two time scales. This decomposition allows to design separately two observers, and then to interconnect them in cascade. The first one works on a bounded time scale, explaining why the system is not identifiable at steady state, while the second one works on unbounded time scale. Finally, this construction provides a practical convergence of the coupled observers. Each observer has been built considering the variable high-gain technique proposed in [10] with an explicit construction of Lipschitz extensions of the dynamics, similarly to the work presented in [19]. Other choices of observers techniques could have been made and applied to this particular structure. We believe that such a decomposition might be applied to other systems of interest, that are not identifiable or observable at steady state.

Acknowledgements

The authors are grateful to D. Dochain, C. Lobry and J. Harmand for useful discussions. The first author acknowledges the financial support of INRA.

Author details

Miled El Hajji
ISSATSO, Université de Sousse, cité taffala, 4003 Sousse, Tunisie and LAMSIN, BP. 37, 1002 Tunis-Belvédère, Tunis, Tunisie

Alain Rapaport
UMR INRA-SupAgro 'MISTEA' and EPI INRA-INRIA 'MODEMIC' 2, place Viala, 34060 Montpellier, France

8. References

[1] J. Baranyi and T. Roberts. A dynamic approach to predicting microbial growth in food. *Int. J. Food Microbiol.* 23, pp. 277–294, 1994.

[2] J. Baranyi and T. Roberts. Mathematics of predictive food microbiology. *Int. J. Food Microbiol.* 26, pp. 199–218, 1995.

[3] J. Baranyi, T. Roberts and P. McClure. A non-autonomous differential equation to model bacterial growth. *Food Microbiol.* 10, pp.43–59, 1993.

[4] E. Beretta, G. Bischi and F. Solimano. Stability in chemostat equations with delayed nutrient recycling, *J. Math. Biol.*, 28 (1), pp. 99–111, 1990.

[5] E. Beretta and Y. Takeuchi. Global stability for chemostat equations with delayed nutrient recycling. *Nonlinear World*, 1, pp. 191–206, 1994.

[6] R. Buchanan. Predictive food microbiology. *Trends Food Sci. Technol.* 4, pp. 6–11, 1993.

[7] G. Ciccarella, M. Dalla Mora, and A. Germani A Luenberger-like oberver for nonlinear systems. *Int. J. Control*, 57 (3), pp. 537–556, 1993.

[8] M. Dalla Mora, A. Germani, and C. Manes A state oberver for nonlinear dynamical systems. *Nonlinear Analysis, Theory, Methods & Applications*, 30 (7), pp. 4485–4496. , 1997.

[9] H. Freedman and Y. Xu. Models of competition in the chemostat with instantaneous and delayed nutrient recycling *J. Math. Biol.*, 31 (5), pp. 513–527, 1993.

[10] J.P. Gauthier, H. Hammouri, and S. Othman A simple observer for nonlinear systems: Applications to bioreactors.*IEEE Transactions on automatic control*, 37 (6), pp. 875–880., 1992.

[11] K. Grijspeerdt and P. Vanrolleghem Estimating the parameters of the Baranyi-model for bacterial growth. *Food Microbiol.*, 16, pp. 593–605, 1999.

[12] X. He and S. Ruan. Global stability in chemostat-type plankton models with delayed nutrient recycling. *J. Math. Biol.* 37 (3), pp. 253–271, 1998.

[13] S. Jang. Dynamics of variable-yield nutrient-phytoplankton-zooplankton models with nutient recycling and self-shading. *J. Math. Biol.* 40 (3), pp. 229–250, 2000.

[14] L. Jiang and Z. Ma. Stability of a chemostat model for a single species with delayed nutrient recycling-case of weak kernel function, *Chinese Quart. J. Math.* 13 (1), pp. 64–69, 1998.

[15] J. Lobry, J. Flandrois, G. Carret, and A. Pavé. Monod's bacterial growth revisited. *Bulletin of Mathematical Biology*, 54 (1), pp. 117-122, 1992.

[16] Z. Lu. Global stability for a chemostat-type model with delayed nutrient recycling *Discrete and Continuous Dynamical Systems - Series B*, 4 (3), pp. 663–670, 2004.

[17] V. Lubenova, I. Simeonov and I. Queinnec Two-step parameter and state estimation of the anaerobic digestion, *Proc. 15th IFAC Word Congress*, Barcelona, July 2002.

[18] S. Pirt. Principles of microbe and cell cultivation. *Blackwell Scientific Publications London*, 1975.

[19] A. Rapaport and A. Maloum. Design of exponential observers for nonlinear systems by embedding. *International Journal of Robust and Nonlinear Control*, 14, pp. 273–288, 2004.

[20] S. Ruan. Persistence and coexistence in zooplankton-phytoplankton-nutrient models with instantaneous nutrient recycling *J. Math. Biol.* 31 (6), pp. 633–654, 1993.

[21] S. Ruan and X. He. Global stability in chemostat-type competition models with nutrient recycling, *SIAM J. Appl. Math.* 58, pp. 170–192, 1998.

[22] K. Rudi, B. Moen, S.M. Dromtorp and A.L. Holck Use of ethidium monoazide and PCR in combination for quantification of viable and dead cells in complex samples, *Applied and Environmental Microbiology*, 71(2), p. 1018–1024, 2005.

[23] I. Simeonov, V. Lubenova and I. Queinnec Parameter and State Estimation of an Anaerobic Digestion of Organic Wastes Model with Addition of Stimulating Substances, *Bioautomation*, 12, pp. 88–105, 2009.

[24] Z. Teng, R. Gao, M. Rehim and K. Wang. Global behaviors of Monod type chemostat model with nutrient recycling and impulsive input, *Journal of Mathematical Chemistry*, 2009 (in press).

[25] S. Yuan, W. Zhang and M. Han. Global asymptotic behavior in chemostat-type competition models with delay *Nonlinear Analysis: Real World Applications*, 10 (3), pp. 1305–1320, 2009.

Chapter 4

Generation and Utilization of Microbial Biomass Hydrolysates in Recovery and Production of Poly(3-hydroxybutyrate)

Jian Yu, Michael Porter and Matt Jaremko

Additional information is available at the end of the chapter

1. Introduction

In moving towards sustainable manufacturing with reduced carbon footprint, bio-based fuels, chemicals and materials produced from renewable resources have attracted great interest. Microbial cells, in working with other chemical and enzymatic catalysts, are often used in the conversion of feedstocks to desired products, involving different species of bacteria, yeast, filamentous fungi, and microalgae. A substantial amount of microbial biomass is generated in the industrial fermentations and often discarded as a waste. Because of a high cost associated with growth and disposal of the cell mass, reusing the microbial biomass may be an attractive alternative to waste disposal. In contrast to the biomass as energy storage (e.g. starch and oil) or plant structure (e.g. cellulose and hemi-cellulose), microbial biomass is biologically active, consisting primarily of proteins (10-60 wt%), nucleic acid (1-30 wt%), and lipids (1-15%) [1]. Few cases of reusing microbial biomass exist in industrial processes.

Poly(3-hydroxybutyrate) (PHB) is a representative polyhydroxyalkanoate (PHA) that is formed by many bacterial species as carbon and energy reserve [2,3]. Although the biopolyesters made from renewable feedstocks have the potential of replacing petroleum-based thermoplastics in many environmentally friendly applications, they are not widely accepted in the markets because of the high production cost [4]. Extensive research has been conducted to use cheap feedstocks [5,6], develop high cell density fermentation technology for high PHA productivity [7,8], and improve microbial strains that exhibit good performance under high osmotic pressure and environmental stress [9-11]. One major cost factor of PHA production is the recovery and purification of biopolyester for desired purity and material properties [4, 12]. Depending on strains and culture conditions, biopolyester may account for 50-80 wt% of cell mass [13]. They are stored in microbial cells as tiny amorphous granules [0.2-0.5 μm in diameter] and need a sophisticated treatment to separate them from the residual cell mass [14-16]. Two technologies, based-on solvent extraction or biomass dissolution, are usually adopted in PHA recovery. With solvent extraction, the PHA granules are dissolved in appropriate organic solvents, leaving cells or residual biomass intact [17,18]. With cell mass dissolution, the PHA granules are left intact while the non-PHA cell mass is decomposed and dissolved in aqueous solutions with help of biological and/or chemical agents [19,20]. Following either treatment, PHA and non-PHA cell mass can be separated with conventional solid/liquid separations. Separating biopolyester from the cells would generate a substantial amount of residual microbial biomass, 0.25 to 1 kg dry mass per kg of PHA resin, depending on the initial PHA content. As a mixture of proteins, nucleic acids, lipids, and wall fragments, the residual biomass has no market value and is discarded at an extra disposal cost.

According to the microbial structure and cellular composition [1], the residual microbial biomass is actually consisting of true biological compounds formed during cell growth while PHA is just a carbon storage material. In a conventional PHA fermentation, sufficient substrates and nutrients (C, N, P, minerals and some organic growth factors) are supplied to grow enough cell mass that in turn or simultaneously synthesize biopolyester from carbon substrates. A large portion of organic carbons and nutrients are therefore consumed to generate new cell mass that is going to be discarded as a solid waste after polymer recovery. Ideally, the residual biomass should be reused by microbial cells to generate new cell mass and/or PHA polymers. This would not only reduce the cost of waste treatment and disposal, but also save the cost of nutrients in PHA fermentation. The bacterial cells, however, cannot assimilate their cell mass because they lack appropriate enzymes to break down various biological macromolecules and their complex structure such as cellular walls and membranes [19]. If the cells or mutants from genetic engineering could easily assimilate their structural components, they might not be suitable to industrial PHA production because the cells would undergo autolysis under high environmental stress. It is highly possible, however, to make the residual biomass reusable during PHA recovery in which the non-PHA cell mass is decomposed and hydrolyzed in aqueous solutions [20]. Integration of PHA recovery with reusing of residual biomass in microbial PHA fermentation is a novel and challenging technology. This work shows the generation of biomass hydrolysates in a downstream PHB recovery and the beneficial utilization of the hydrolysates in cell growth and PHB formation.

2. Materials and methods

2.1. Microbial cultures

A laboratory strain of *Ralstonia eutropha* maintained on agar slant was used in this work. The agar medium contained (per liter): 5 g yeast extract, 5 g peptone, 2.5 g meat extract and 15 g agar. A 16S RNA analysis indicates that the strain has 100% alignment with *Ralstonia eutropha* H16. The cells were cultivated in 200 mL mineral medium containing: 160 mL glucose mineral solution, 8 mL inoculum, 5-20 mL solutions of residual microbial biomass hydrolysates, and the rest pre-sterilized water. The glucose mineral solution contained (per liter): 12 g glucose, 2 g NaH$_2$PO$_4$, 2.8 g K$_2$HPO$_4$, 0.5 g MgSO$_4$.7H$_2$O, 1 g (NH$_4$)$_2$SO$_4$ and 1 mL trace solution [21]. The culture solution was shaken in a 500 mL baffled flask at 30 $^{\circ}$C and 200 rpm in a rotary incubator for 24 or 48 hours. Cell mass was harvested from 50 mL medium with centrifugation at 5,000 g for 10 min and freeze dried for measurement of cell mass concentration and PHB content.

A large amount of PHB-containing cell mass for biopolyester recovery was produced with a fed-batch culture in a 3L bench-top bioreactor (BioFlo 110, New Brunswick Scientific Co. NJ). The temperature, pH and dissolved oxygen were controlled at 30 $^{\circ}$C, 6.8, and 10% air situation, respectively. A feed solution prepared with a sugar manufacturing byproduct containing about 50% sucrose was introduced into the bioreactor till the cell density reached 130 g/L and PHB concentration 94 g/L (72% PHB of cell mass). The cell mass was harvested with centrifugation and re-suspended in an acidic water (0.2M H$_2$SO$_4$) to make a cell slurry of 278 g dry mass/L. The slurry was reserved for later use.

2.2. PHB recovery

The biopolyester was recovered and purified by dissolving the non-PHA cell mass (28% w/w) in sequential treatments consisting of acid pretreatment, base treatment, hypochlorite whitening, washing and drying [22].

Acid pretreatment: The cell mass in acidic solution was heated to boil and maintained for one hour under ambient conditions. The pretreated cellular solids were cooled to room temperature and separated from acid solution with centrifugation at 5,000 g for 10 min. The dissolved microbial biomass in the supernatant solution is referred to acid hydrolysates. The insoluble wet pellets were subjected to next base treatment.

Base treatment: The insoluble solids from the acid pretreatment were re-suspended in an equivalent volume of water to form slurry of about 200 g DM/L. The solution pH was raised to 10 to 11 with 5 M NaOH solution and stirred under ambient conditions for 30 to 60 min. A small amount of surfactants such as sodium dodecylsulfate (SDS, CH$_3$(CH$_2$)$_{11}$OSO$_3$Na) might be added to a concentration of 5 to 10 g/L. The slurry was then heated to boiling and maintained for 10 min under ambient conditions. After centrifugation, the dissolved biomass in the supernatant solution is referred to base hydrolysates. The sequential acid and base treatments above could also be performed without separation of the acid hydrolysates. In this case, sodium hydroxide was directly added into the acidic cell slurry and the pH was raised to 10-11 for base treatment. After centrifugation, the supernatant solution contained the hydrolysates generated from both acid and base treatments and is refereed to acid-base hydrolysates.

Whitening and washing: The insoluble wet pellets from base treatment were re-suspended in a commercially available bleaching solution containing 6% w/w of hypochlorite. The slurry was stirred for 1 to 2 hours under ambient conditions. The white PHB pellets were recovered with centrifugation. A small amount of biomass was dissolved and mineralized in the bleaching solution because of chemical oxidation, which is not considered for reuse in this work. The wet PHA pellets were washed two times with water and dried in oven. The final PHB product is a white powder.

2.3. Chemical and material analysis

The dissolution of non-PHA cell mass was monitored by measuring the characteristic absorption of amino acid residues at 280 nm with a UV/VIS spectrophotometer (Beckman Coulter DU530, Fullerton, CA). The concentration of proteins that can be stained in Bradford assay was measured with the spectrophotometer after protein-dye binding [23]. The content of PHB in original cell mass, in sequential treatments, and final product were determined via methanolysis of the biopolyester in methanol (3 wt% sulfuric acid) at 100 °C for 8-10 hours [24]. The 3-hydroxybutyric methyl ester was hydrolyzed into 3-hydroxybutyric acid when the solution pH was raised to 11 with a 10N NaOH solution. The liquid samples were analyzed using an HPLC equipped with a UV detector (Shimadzu, Japan) and an organic acid column (OA-1000, Alltech, Deerfield, IL). The column was maintained at 65 °C and eluted with a water-sulfuric acid solution (pH 2) at 0.8 mL/min. The monomeric acid and crotonic acid, a trace byproduct formed in methanolysis, were detected at 210 nm. For data quality control, the biopolyester was also extracted from the freeze-dried cell mass in hot chloroform followed by precipitation with methanol [21]. The PHB content was calculated from the purified PHB and compared with the results of HPLC analysis.

The purified PHB and non-PHB cell mass were examined with a Nicolet Avatar 370 FTIR spectrometer (Thermo Electron Co., Madison, WI). The solids were pressed on a germanium crystal window of micro-horizontal attenuated total reflectance (ATR) for measurement of single-reflection and absorption of infrared radiation by the specimens. The thermal properties of PHB powder were examined with a differential scanning calorimeter (DSC). A Modulated 2920 instrument (TA Instruments, New Castle, DE) equipped with a refrigerated cooling system was run in heat-cool-heat mode at a rate of 5 °C/min under nitrogen. The selected temperature range was 30°C – 210 °C with sample weights of 4.5 – 5.5 mgs. Images of cell and PHB granules were obtained with an energy-filtering transmission electron microscopy (120 kV LEO 912, Carl Zeiss SMT Inc. MA). The instrument has an in-column electron energy loss spectrometer, allowing analysis of light element in thin sections.

3. Results and discussion

3.1. Sequential treatment for PHB recovery

Figure 1 elucidates the sequential treatment of PHB-containing cells in a process of PHB recovery and purification. Starting with 100 dry cell mass, the cells in a slurry of 278 g DM/L were first treated in an acidic solution (0.2M H_2SO_4). A substantial amount of microbial proteins was released from the damaged cells, depending on temperature and time as shown in Figure 2.

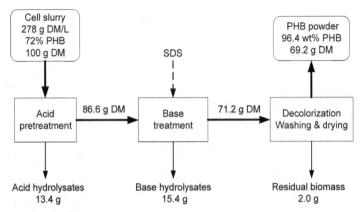

Figure 1. Sequential treatment of PHB-containing biomass (100 g dry mass) and generation of biomass
hydrolysates. Surfactant SDS is optional for high PHB purity.

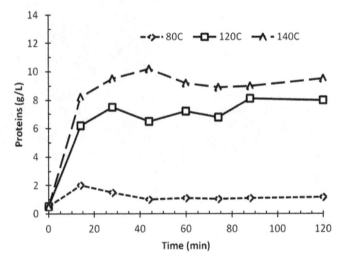

Figure 2. Effect of temperature and time on release of proteins from microbial cells in acidic solution.

During the acid pretreatment, the original amorphous PHB granules became partially
crystallized (data not shown here), which improved the granule's resistance to abiotic
degradation in the following treatments [20]. The biomass hydrolysates (13.4 g dry mass)
dissolved in supernatant solution was discharged as acid hydrolysates. The residual PHB-
containing biomass was further subjected to a base treatment by raising the slurry pH to
10.5 with a 10M NaOH solution. About 15.4 g dry mass was dissolved in the supernatant
solution and discharged as base hydrolysates. After a small amount of residual biomass (2 g
dry mass) was removed via oxidation with hypochlorite, the final PHB powder (69.2 g dry
mass) contained 96.4 wt% PHB. The overall PHB recovery yield was 92.6%, or 7-8% PHB

was lost in repeated hydrolysis and solid/liquid separations. When a small amount of surfactant such as sodium dodecylsulfate (SDS) was added in the base treatment, the PHB purity of the final biopolyester resin could be increased to 99.4%.

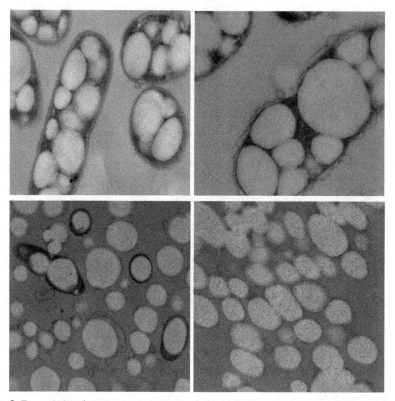

Figure 3. Transmission electron microscope images: microbial cells containing native PHB granules (top left), cells with damaged walls in acid pretreatment (top right), PHB granules with attached residual cell mass (bottom left), and purified PHB granules (bottom right)

Figure 3 is the electronic microscopy images of the original cells with PHB inclusion bodies, the cells with damaged porous cell walls in acid pretreatment, the PHB granules with residual cellular mass in base treatment, and the purified PHB granules after whitening and washing. It is interesting to see that the cell walls became porous in the acid pretreatment, which allowed release of proteins and other biological components in cytoplasm. The original cell structure, however, was maintained to keep the PHB granules within the damaged cells. After base treatment, the cell walls were almost completely decomposed, and a small amount of residual cell mass, probably some hydrophobic cellular components, was attached to PHB granules. After whitening and washing, the non-PHA cell mass was removed to give purified PHB granules.

3.2. Purified PHB versus original cell mass

Figure 4 compares the FTIR spectra of the purified PHB granules and the original oven-dried PHB-containing cell mass. The peaks of amide I band at ~1650 cm⁻¹ and amide II band at ~1540 cm⁻¹ (N-H bend) are characteristic infrared radiation absorption of the proteins in cell mass [25,26]. They disappeared in the spectrum of purified PHB granules. This was confirmed with a pure PHB prepared with solvent extraction (the spectrum not shown here). It was also noticed that the native amorphous PHB granules became crystallized during the process of purification, which will be further discussed. This structural change in PHB matrix was also reflected in the absorption of infrared radiation at wave numbers of 1180, 1210, and 1280 nm⁻¹ [27].

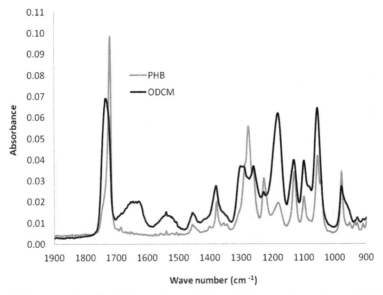

Figure 4. FTIR spectra of purified PHB granules and PHB-containing oven-dried cell mass (ODCM).

The purified PHB granules were subjected to repeated heating and cooling in a differential scanning calorimeter (DSC), and the results are presented in Figure 5. In the first heating (solid blue line), two melting peaks were observed, around 156 °C and 167 °C, respectively, indicating that the PHB powder had two types of crystalline structures. The melted polymer was re-crystallized again during the first cooling (dotted blue line), starting at 100 °C and ending at 80 °C. When the crystallized PHB was subjected to the second heating (solid black line), it was noticed that the relatively small melting peak at 156 °C (or crystalline structure) observed in the first heating disappeared. Only one melting peak (crystalline structure) was observed, starting at 160 °C and ending at 181 °C with a peak around 174 °C. This phenomenon indicates that the first melting peak in the first heating represents a type of crystalline structure that could not be formed at a cooling rate of 5 °C/min. The whole endothermic event in the second heating absorbed 87J/g PHB. Based on a theoretical melting enthalpy of 100% crystalline PHB (146 J/g) [28], it can be estimated that about 60% of PHB matrix was crystallized during the first cooling (Eq. 1).

$$X_C = \frac{\Delta H_m}{\Delta H_t} = \frac{87}{146} \approx 60\% \tag{1}$$

Where X_c is the PHB crystallinity, ΔH_m and ΔH_t are the melting enthalpies of PHB powder and a theoretical PHB crystal [28].

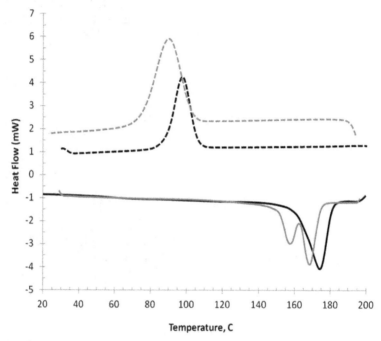

Figure 5. Differential scanning calorimetry (DSC) measurement of purified PHB granules in repeated heating and cooling: the first heating (solid blue line) followed by the first cooling (dotted blue line) and the second heating (solid black line) following by the second cooling (dotted black line).

In contrast to the thermal plastic behavior of PHB powders, the oven-dried cell mass containing 73% PHB could not be melted till carbonization. This fact reveals a complicated interaction between the biopolyester and the residual biomass. It also shows that a composite of 73% PHB and 27% cellular mass is not a thermoplastic material, but a rigid composite. The role of cellular mass in the PHB composite is not clear yet.

3.3. Cell debris solutions

As shown in Figure 1, about 94 wt% of residual microbial biomass was decomposed and hydrolyzed in the acid pretreatment (~44% biomass) and base treatment (~50% biomass). The remaining small amount (~6 wt %) of residual cell mass is most likely mineralized via oxidation with hypochlorite, a strong oxidation agent. The acid hydrolysates are primarily the cytoplasm proteins released from the damaged cells (Figures 2 and 3). The released biological macromolecules were subjected to further hydrolysis in the thermal acidic solution. The acid hydrolysates solution had a clean brownish color and contained 30-45 g/ of soluble biomass, depending on the density of cell slurry and treatment conditions. The base hydrolysates solution with a dark color contained the hydrolysis products of hydrophobic cell components including lipids and membrane proteins. After centrifugation, the concentration of soluble biomass in the supernatant solution was 30 to 50 g/L. The sequential treatments disrupted and dissolved the structural components so that they could be removed from PHB granules. Equally important, the biomass and biological macromolecules were decomposed into small molecule hydrolysates such as amino acids and organic acids. These hydrolysates could become appropriate substrates that can be assimilated by microbial cells in microbial PHB production.

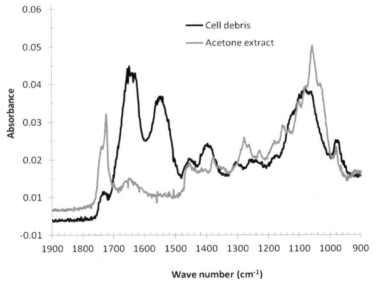

Figure 6. FTIR spectra of acid-base biomass hydrolysates (black line, cell debris) and cellular components extracted with acetone (blue line)

In addition to the two types of biomass hydrolysates described above, a mixed hydrolysates of residual biomass was generated when the acid pretreatment and base treatment were performed sequentially without solid/liquid separation. It eliminated one operation of solid/liquid separation, but the acid hydrolysates (primarily proteins) were subjected to additional hydrolysis in the base solution. Figure 6 shows the FTIR spectrum of an acid-base hydrolysates. As observed in the IR spectrum of the original cell mass in Figure 4, a major component of the hydrolysates was the amino acids or proteins with infrared radiation absorbance at 1500 to 1700 cm^{-1} [25]. Another major component in the acid/base hydrolysates had the IR absorbance at 1000 to 1200 cm^{-1}, which was attributed to cell lipids and/or similar compounds. This was confirmed with the spectrum of cell mass extract in acetone. Acetone is a common solvent used to remove hydrophobic lipids, steroids and pigments from PHB-containing cell mass [17]. It does not dissolve and extract PHB and proteins. Based on the observations above, it was concluded that the major cellular components in the acid hydrolysates were derived from cytoplasm proteins and in the base hydrolysates from cell walls, lipids and membrane proteins. In the acid-base hydrolysates, the products were derived from both groups, i.e. amino acids or peptides derived from proteins, and lipids derived from cell walls and membranes. It should be pointed out that the composition of acid-base hydrolysates is not a simple mixture of acid- and base-hydrolysates because the acid hydrolysates were further hydrolyzed in base treatment.

Because of the hydrophobic properties of PHB granules, the residual hydrophobic impurities of cell mass might be attached to the granules and difficult to remove by washing with water. A surfactant such as SDS in the base treatment can remove most of the impurities to a high PHB purity (>99 % w/w). A large portion of the surfactant, however, may be left in the hydrolysates solution and may have an adverse effect on the reuse of the hydrolysates in PHB production.

3.4. Utilization of acid hydrolysates in PHB biosynthesis

An acid hydrolysates solution containing 38 g/L of soluble solids was added into a glucose medium to give a predetermined percentage of residual biomass to glucose at 0, 10, 20 and 25% of sugar, respectively. The initial glucose concentration was controlled at a constant level of 9.6 g/L. The flask cultures of no biomass hydrolysates were run in parallel as controls. As shown in Figure 7, the acid biomass hydrolysates were beneficial to both cell growth and PHA formation. Because the residual biomass might also contain some insoluble solids and PHB granules lost in PHB recovery, both cell density and PHB concentration were compared at 24 hours and 48 hours to show the net gains. The benefits of biomass hydrolysates were statistically significant based on the deviations of duplicates.

The acid hydrolysates might have two positive effects on microbial PHA formation. First, the hydrolysates promoted cell activity on glucose utilization, giving higher cell densities than the controls in the first 24 hours. This nutritional effect was similar to those of organic nutrients such as yeast extract and peptone, which are widely used in microbial cultures to provide nutrients and growth factors to the cells. A fast cell growth can reduce the cultivation time, resulting in a high PHB productivity. Second, the biomass hydrolysates might also be used as an extra carbon source to generate more cell mass than the controls in 48 hours. This carbon source effect, however, might play a minor role because the cell density did not increase with cell debris load. In fact, too much acid hydrolysates deteriorated the gains as shown in Figure 7. The reason is not clear yet. A load of acid biomass hydrolysates to glucose from 10 to 20 wt% seems appropriate for both cell growth and PHB formation.

3.5. Utilization of acid-base biomass hydrolysates

An acid-base biomass hydrolysates solution containing 48.5 g/L of soluble solids was added into a glucose medium at predetermined percentage of cell debris to glucose from 0 to 40%. The glucose medium without biomass hydrolysates was run in parallel as controls. As shown

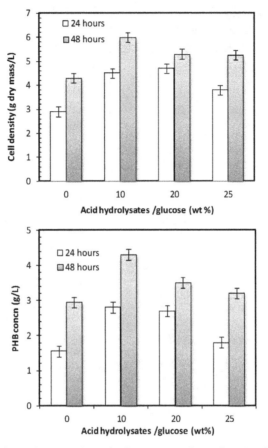

Figure 7. Effect of acid hydrolysates to glucose loading ratio on cell growth (top) and PHB formation (bottom) in reuse of the residual biomass for PHB production.

in Table 1, the concentrations of both cell mass and PHA content, after 48 hours cultivation, were substantially higher than those of the control. The overall cell growth yield (Yx/s) and PHA formation yield (Yp/s) are calculated from the amounts of cell mass and PHA formed in 48 hours based on the initial concentration of glucose. The relative yields (Yx' and Yp') based on the controls were increased by 100 – 300%. More interestingly, the inhibitory effect of acid biomass hydrolysates was not observed in the use of acid-base hydrolysates, and the load of cell debris to glucose can be increased to about 39 wt%. Most likely, the unknown inhibitors in acid hydrolysates were further decomposed into less inhibitory hydrolysates in the base treatment. Since the acid biomass hydrolysates contained primarily the cytoplasm proteins released from the damaged cells, the soluble proteins might adversely affect cell growth at high concentrations. After being hydrolyzed in base solution into peptides and amino acids, the small molecule hydrolysates become less inhibitory and more usable to the cells.

Biomass hydrolysates (g/L)	Biomass /Glucose (wt%)	Cell density (g/L)	PHB (wt%)	Yx/s (g/g)	Yp/s (g/g)	Yx′	Yp′
0.0	0	2.1 ± 0.5	45 ± 1.2	0.18	0.081	1.0	1.0
1.15	9.4	4.0 ± 0.2	55 ± 1.6	0.33	0.18	1.8	2.24
2.30	18.8	4.5 ± 0.3	60 ± 1.7	0.38	0.23	2.1	2.84
4.60	38.8	5.6 ± 0.4	61± 1.5	0.47	0.29	2.6	3.58

Note: the flask cultures were maintained at 30 °C and 200 rpm in a rotary incubator for 48 hours. The yields of cell mass (Yx/s) and PHB (Yp/s) were based on the initial glucose concentration. The relative yields (Yx′ and Yp′) are the ratios of yields with hydrolysates to the control without hydrolsyates.

Table 1. Effect of acid-base biomass hydrolysates on cell growth and PHB formation

3.6. Comparison of three types of cell debris

In the process of PHB recovery (Figure 1), three types of biomass hydrolysates may be generated, depending on operations: acid, base, and acid-base biomass hydrolysates. They may have different nutrient values or inhibitory effects on cell growth and PHB synthesis. Solutions of three types of biomass hydrolysates were added into a glucose medium for pre-determined concentrations of cell debris. Controls without hydrolysates were run in parallel. The ratios of cell densities (g/L) to the controls were compared after 48 hours cultivation as shown in Figure 8. The nutritional value of acid hydrolysates in cell growth is similar to that of base hydrolysates. The nutrient value of acid-base hydrolysates, however, is significantly higher than those of hydrolysates from individual treatment.

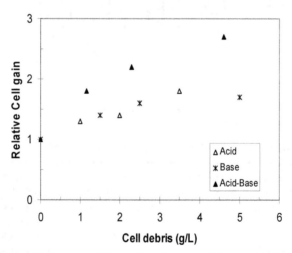

Figure 8. Comparison of three types of biomass hydrolysates (acid, base, and acid-base) on cell growth in a glucose mineral medium. The relative cell gain is the ratio of cell density to the controls.

Based on an average cell yield (Yx/s = 0.45) of PHB fermentation on glucose [29], 45 kg of cell mass containing 70 wt% of PHB is generated from 100 kg of glucose consumed. A downstream recovery and purification as shown in Figure 1 can generate 31 kg PHB resin and 13 kg acid-base hydrolysis of residual microbial biomass. It is assumed that the acid hydrolysates are not separated, but hydrolyzed sequentially in the base treatment and discharged with the base hydrolysates together. If this amount of residual biomass is reused in next PHB fermentation, the percentage of biomass hydrolysates to glucose is 13% at maximum (13 kg for 100 kg glucose), a moderate load of biomass hydrolysates (Table 1). In real fermentations, more glucose is often added because of the residual glucose in the spent medium. This quick calculation indicates that most of residual biomass discharged from downstream separations can be reused in the next PHA fermentation. In addition to the elimination of a waste stream, the productivity and yields of PHA fermentation can also be significantly improved.

3.7. Effect of SDS in biomass hydrolysates solution

SDS is a popular surfactant used in PHA recovery to disrupt the cells or remove a small amount of hydrophobic impurities from PHB granules [30, 31]. The purity of PHB granules can be increased from 96.4% to above 99% when a small amount of SDS was added in the base treatment. The surfactant left in the base solution, however, may have an adverse effect when the cell debris is reused in microbial PHB fermentation. An acid-base biomass hydrolysates solution containing 48.5 g/L of soluble solids and SDS were added into a glucose medium. The cell debris concentration was kept at 1.94 g/L, and SDS concentration was increased from 0.2 to 0.8 g/L with SDS. Controls without biomass hydrolysates and SDS were run in parallel. Table 2 gives the results of cell growth and PHA formation at different surfactant levels at 24 and 48 hours, respectively.

Biomass hydrolysates (g/L)	SDS (g/L)	24 hours		48 hours	
		Cell mass (g/L)	PHB (wt%)	Cell mass (g/L)	PHB (wt%)
0	0	2.08 ± 0.02	36.8 ± 1.1	2.7 ± 0.02	44.4 ± 0.6
1.94	0.2	3.69 ± 0.06	49.7 ± 2.1	4.99 ± 0.04	60.6 ± 1.1
1.94	0.4	3.27 ± 0.05	39.6 ± 1.2	4.58 ± 0.04	53.8 ± 0.8
1.94	0.6	2.70 ± 0.04	33.9 ± 1.5	3.51 ± 0.03	50.3 ± 0.7
1.94	0.8	2.59 ± 0.04	17.8 ± 1.3	3.54 ± 0.03	25.8 ± 1.2

Note: flask cultures were maintained at 30 ℃ and 200 rpm in a rotary incubator.

Table 2. Effect of surfactant SDS and biomass hydrolysates on cell growth and PHB formation

Compared with the controls of no biomass hydrolysates and surfactant, all the cultures containing the acid-base hydrolysates exhibited better cell growth. Particularly, the increase of cell concentration from 24 hours to 48 hours was the new cell mass formed in 24 hours from glucose and cell debris in the presence of SDS. The formation of PHB, however, was deteriorated at high SDS concentrations (0.6-0.8 g/L). At a low or moderate SDS concentrations (0.2-0.4 g/L), the positive effect from biomass hydrolysates was much higher than the negative effect of surfactant. The PHB concentrations, after 48 hours cultivation, were 2.4 to 3 g/l, in comparison with 1.2 g/L of the control. The results in Table 2 indicate that the dosage of SDS in PHA recovery should be controlled according to the amount of residual microbial biomass generated. The mass ratio of SDS to biomass hydrolysates should be less than 20% w/w or better at 10% w/w. In a typical PHB recovery process as shown in Figure 1, the amount of SDS used should be less than 2.9 g for 10% of acid-base cell debris or 5.8 g for 20% of acid-base cell debris. The consumption of SDS is therefore 4-8 % of PHB resin produced. It is much lower than the SDS dosages used in the conventional separations [30, 31].

4. Conclusion

Residual microbial biomass is an inevitable waste generated in downstream recovery of polyhydroxyalkanoates from microbial cells. With a separation technology based on sequential dissolution of no-PHB cell mass in aqueous solutions, the cell mass separated from the PHB-granules is decomposed and hydrolyzed into small molecule hydrolysates that can be assimilated by microbial cells as nutrients and/or carbon source. A type of biomass hydrolysates generated from continuing treatment in acid and base solutions exhibits the best nutrient value for cell growth and PHA formation. The acid-base hydrolysates contains two major water-soluble components derived from the cell proteins and lipids, respectively. When PHB-producing cells are fed with the hydrolysates in a glucose mineral solution, the cells grow faster and form more biopolyester in comparison with the controls that do not contain the hydrolysates. The glucose-based yields of cell mass and PHA bioplastics are significantly improved. SDS is an efficient surfactant to remove the small amount of hydrophobic residues for high PHB purity, but also a potential inhibitor to microbial PHA formation. When the amount of surfactant is less than 20% of an acid-base biomass hydrolysates, its negative effect is overwhelmed by the nutritional value of hydrolysates. Under these conditions, it is highly possible to reuse most of the residual biomass discharged from PHB recovery in the next microbial PHB fermentation. It therefore eliminates a waste stream from bioplastics production and saves the nutrients with improved PHA productivity and yield.

Author details

Jian Yu, Michael Porter and Matt Jaremko
Hawaii Natural Energy Institute, University of Hawaii at Manoa, Honolulu, Hawaii, USA

Acknowledgement

The authors acknowledge a support from Bio-On to this work. MP and MJ are graduate students of Molecular Bioscience & Bioengineering at UHM.

5. References

[1] Shuler ML, Kargi F (1992) Bioprocess Engineering: Basic Concepts. New Jersey: Prentice-Hall Inc. pp. 48-50.

[2] Steinbuchel A, Valentin HE (1995) Diversity of bacterial polyhydroxyalkanoic acids. *FEMS Microbiol. Lett.* 128:219-228.

[3] Lenz RW, Marchessault RH (2005) Bacterial polyesters: biosynthesis, biodegradable plastics and biotechnology. *Biomacromolecules* 6:1-8.

[4] Yu J (2006) Microbial production of bioplastics from renewable resources. In: Yang ST, editor. Bioprocessing of value added products from renewable resources. Amsterdam: Elsevier. pp. 585-610.

[5] Solaiman DKY, Ashby RD, Foglia TA, Marmer WN (2006) Conversion of agricultural feedstock and coproducts into poly(hydroxyalkanoates). *Appl. Microbiol. Biotechnol.* 71:783-789.

[6] Wu Q, Huang H, Hu G, Chen J, Ho KP, Chen GQ (2001) Production of poly-3-hydroxybutyrae by *Bacillus* sp. JMa5 cultivated in molasses media. *Antonie van Leeuwenhoek* 80:111-118.

[7] Park SJ, Park JP, Lee SY (2002) Production of poly(3-hydroxybutyrate) from whey by fed-batch culture of recombinant *Escherichia coli* in a pilot-scale fermenter. *Biotechnol. Lett.* 24:185-189.

[8] Madden LA, Anderson AJ, Asrar J (1998) Synthesis and characterization of poly (3-hydroxybutyrate) and poly(3-hydrybutyrate-co-3-hydroxyvalerate) polymer mixtures produced in high-density fed-batch cultures of *Ralstonia eutropha*. *Macromolecules* 31:5660-5667.

[9] Full TD, Jung DO, Madigan MT (2006) Production of poly-β-hydroxyalkanoates from soy molasses oligosaccharides by new, rapidly growing *Bacillus* species. *Lett. Appl. Microbiol.* 43:377-384.

[10] Tsuge T (2002) Metabolic improvements and use of inexpensive carbon sources in microbial production of polyhydroxyalkanoates. *J. Biosci. Bioeng.* 94:579-584.

[11] Zhang H, Obias V, Gonyer K, Dennis D (1994) Production of polyhydroxyalkanoates in sucrose-utilizing recombinant *Esherichia coli* and *Klebsiella* strains. *Appl. Environ. Microbiol.* 60:1198-1205.

[12] Acquel N, Lo CW, Wei YU, Wu HS, Wang SS (2008) Isolation and purification of bacterial poly(3-hydroxyalkaoates). *Biochem. Eng. J.* 39:15-27.

[13] Brandl H, Gross RA, Lenz RW, Fuller RC (1990). Plastics from bacteria and for bacteria: Poly(β-hydroxyalkanoates) as natural, biocompatible, and biodegradable polyesters. *Adv. Biochem. Eng. Biot.* 41:77-93.

[14] McCool GJ, Fernandez T, Li N, Cannon MC (1996) Polyhydroxyalkanoate inclusion-body growth and proliferation in *Bacillus megaterium*. *FEMS Microbiol. Lett.* 138:41-48.

[15] Stuart ES, Tehrani A, Valentin HE, Dennis D, Lenz RW, Fuller RC (1988) Protein organization on the PHA inclusion cytoplasmic boundary. *J. Biotechnol.* 64:137-144.

[16] de Koning GJM, Lemstra PJ (1992) The amorphous state of bacterial poly[(R)-3-hydroxyalkanoate] in vivo. *Polymer* 33:3292-3294.

[17] Gorenflo V, Schmack G, Vogel R, Steinbuchel A (2001) Development of a process for the biotechnological large-scale production of 4-hydroxyvalerate-containing polyesters and characterization of their physical and mechanical properties. *Biomacromolecules* 2: 45-57.

[18] Fiorese ML, Freitas F, Pais J, Ramos AM, de Aragao GMF, Reis MAM (2009) Recovery of polyhydroxybutyrate (PHB) from *Cuprivavidus necator* biomass by solvent extraction with 1,2-propylene carbonate. *Eng. Life Sci.* 9:454-461.

[19] Kapritchkoff FM, Viotti AP, Alli RCP, Succolo M, Pradella JGC, Maiorano AE, Miranda EA, Bonomi A (2006) Enzymatic recovery and purification of polyhydroxybutyrate produced by *Ralstonia eutropha*. *J. Biotechnol.* 122:453-462.

[20] Yu J, Chen L (2006). Cost effective recovery and purification of polyhydroxyalkanoates by selective dissolution of cell mass. *Biotechnol. Progr.* 22:547-553.

[21] Yu J, Plackett D, Chen LXL (2005). Kinetics and mechanism of the monomeric products from abiotic hydrolysis of poly[(R)-3-hydroxybutyrate] under acidic and alkaline conditions. *Polym. Degrad. Stabil.* 89:289-299.

[22] Yu, J (2009). Recovery and purification of polyhydroxyalkanoates from PHA-containing cell mass. *US Patent 7,514,525*.

[23] Bradford M (1976) A rapid and sensitive method for the quantization of microgram quantities of protein utilizing the principle of protein-dye binding. *Anal. Biochem.* 72:248-254.

[24] Hesselmann RPX, Fleischmann T, Hany R, Zehnder AJB (1999) Determination of polyhydroxyalkanoates in activated sludge by ion chromatographic and enzymatic methods. *J. Microbiol. Methods* 35:111-119.

[25] Chittur KK (1998). FTIR/ATR for protein adsorption to biomaterial surfaces. *Biomaterials* 19:357-369.

[26] Kansiz M, Billman-Jacobe H, McNaughton D (2000) Quantitative determination of the biodegradable polymer poly(β-hydroxybutyrate) in a recombinant *Escherichia coli* strain by use of mid-infrared spectroscopy and multivariate statistics. *Appl. Environ. Microbiol.* 66:3415-3420.

[27] Bloembergen S, Holden DA, Hamer GK, Bluhm TL, Marchessault RM (1986). Studies of composition and crystallinity of bacterial poly(β-hydroxybutyrate-co-β-hydroxyvalerate). *Macromolecules* 19:2865-2871.

[28] Barham PJ, Keller A, Otun EL, Holmes PA (1984) Crystallization and morphology of a bacterial thermoplastic – Poly-3-hydroxybutyrate. *J. Mater Sci.* 19:2781-2794.

[29] Kim BS, Lee SC, Lee SY, Chang HN, Chang YK, Woo SI (1994). Production of poly(3-hydroxybutyraic acid) by fed-batch culture of *Alcaligenes eutrophus* with glucose concentration control. *Biotech. Bioeng.* 892-898.

[30] Chen Y, Xu Q, Yang H, Gu G (2001). Effects of cell fermentation time and biomass drying strategies on the recovery of poly-3-hydroxyalkanoates from *Alcaligenes eutrophus* using a surfactant-chelate aqueous system. *Process Biochem.* 36:773-779.

[31] Strazzullo G, Gambacorta A, Vella FM, Immirzi B, Romano I, Calandrelli V, Nicolaus B, Lama L (2008). Chemical-physical characterization of polyhydroxyalkanoates recovery by means of a simplified method from cultures of *Halomonas campaniensis*. *World J. Microb. Biot.* 24:1513-1519.

Chapter 5

Considerations for Sustainable Biomass Production in *Quercus*-Dominated Forest Ecosystems

Viktor J. Bruckman, Shuai Yan, Eduard Hochbichler and Gerhard Glatzel

Additional information is available at the end of the chapter

1. Introduction

Our current energy system is mainly based on carbon (C) intensive metabolisms, resulting in great effects on the earth's biosphere. The majority of the energy sources are fossil (crude oil, coal, natural gas) and release CO_2 in the combustion (oxidation) process which takes place during utilization of the energy. C released to the atmosphere was once sequestered by biomass over a time span of millions of years and is now being released back into the atmosphere within a period of just decades. Fossil energy is relatively cheap and has been fuelling the world economy since the industrial revolution. To date, fossil fuel emissions are still increasing despite a slight decrease in 2009 as a consequence of the world's economic crisis. Recently, the increase is driven by emerging economies, from the production and international trade of goods and services [1]."If we don't change direction soon, we'll end up where we're heading" is the headline of the first paragraph in the executive summary of the World Energy Outlook 2011 [2]. It unfortunately represents systematic failure in combating climate change and the emphatic introduction of a "green society", leaving the fossil age behind. Certainly such far reaching transformations would take time, but recovery of the world economy since 2009, although uneven, again resulted in rising global primary energy demands [2]. It seems that more or less ambitious goals for climate change prevention are only resolved in phases of a relatively stable economy. Atmospheric carbon dioxide (CO_2) is the second most important greenhouse warming agent after water vapour, corresponding to 26% and 60% of radiative forcing, respectively [3]. Together with other greenhouse gases (GHG's) (e.g. methane (CH_4), nitrous oxide (N_2O) or ozone (O_3)) they contribute to anthropogenic global warming. The industrialization has been driven by fossil sources of energy, emerging in the 17th and 18th century in England as a historical singularity, but soon spreading globally.

Today, our economies still rely on relatively cheap sources of fossil energy, mainly crude oil and natural gas, and consequently emitting as much as 10 PG C per year in 2010 [4]. The Mauna Loa Observatory in Hawaii carries out the most comprehensive and longest continuous monitoring of atmospheric CO_2 concentration. It publishes the well-known Keeling Curve, representing the dynamic change since 1958. Observing the Keeling Curve, one can easily recognize the seasonal variability which is directly triggered by CO_2 uptake of vegetation (biomass) in the northern hemisphere during the vegetation period and secondly, which is even more important in terms of global change, a steady increase of CO_2 concentration from 315 ppmv in 1958 to 394 ppmv in March 2012 [5]. Earlier concentrations could still be derived from air occluded in ice cores. Neftel et al. [6] presents accurate gas concentration measurements for the past two centuries. However, the theoretical knowledge of the warming potential of CO_2 in the atmosphere evolved in the late 19th century when a theory of climate change was proposed by Plass [7], pointing out the "influence of man's activities on climate" as well as the CO_2 exchange between oceans and atmosphere and subsequent acidification. He highlighted the radiative flux controlled by CO_2 in the 12 to 18 micron frequency interval, agreeing with a number of studies published in the forthcoming decades, e.g. Kiehl and Trenberth [3]. In order to understand the fate of anthropogenic CO_2 emissions, research soon focussed on estimating sources and sinks as well as their stability, since it was obvious that the atmospheric concentrations did not rise at the same magnitude as emissions. Available numbers on current fluxes are principally based on the work of Canadell et al., [8] and Le Quéré et al., [1]. In their studies, it is emphasized that the efficiency of the sinks of anthropogenic C is expected to decrease. Sink regions (of ocean and land) could have weakened, source regions could have intensified or sink regions could have transitioned to sources [8]. Another explanation might be the fact that the atmospheric CO_2 concentration is increasing at a higher rate than the sequestration rate of sinks [1]. Moreover, CO_2 fertilization on land is limited as the positive effect levels off and the carbonate concentration which buffers CO_2 in the ocean steadily decreases according to Denman, K.L. et al. [1]. Fossil fuel combustion and land use change (LUC) are the major sources for anthropogenic C emissions (Figure 1). Land use change is usually associated with agricultural practices and intensified agriculture triggers deforestation in developing countries [9] and consequently causes additional emissions.

Another consideration is the availability of fossil fuel, which is limited by the fact that it is a non-renewable and therefore finite resource. Since the fossil energy system is based on globally traded sources with centralized structures, the vulnerability to disturbance is high. Recent examples of price fluctuations caused by political crises or other conflicts in producing countries or along transportation lines demonstrate potential risks. Moreover, a shift towards alternative energy sources and a decentralization of the energy system may contribute to system resilience and create domestic jobs. It prevents capital outflow to unstable political regimes and it helps to protect the environment not only by reducing GHG emissions, but also by reducing impacts of questionable methods of extracting fossil sources of energy (e.g. tar sands exploitation, fracking etc.).

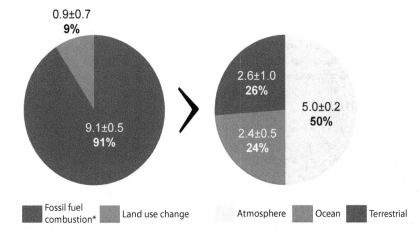

Figure 1. The fate of anthropogenic CO_2 emissions in 2010, showing sources (left) and sinks (right). Presented numbers are Pg C yr^{-1}. The values for 2010 were presented at the Planet under Pressure 2012 conference in London [4].
*includes cement production and flaring.

Biomass could play a significant role in the renewable energy mix. It is a feedstock for bioenergy production and currently thermal utilization (combustion) is by far the most important conversion process, but research activities are focussed on a range of different processes. This includes, for instance, the Fischer-Tropsch synthesis where any kind of biomass may be used as feedstock to produce liquid biofuels. This process is known as biomass – to – liquid (BtL). Research is pushed by national and international regulations (e.g. the EU's 2020 bioenergy target) and commitments, as a climate change mitigation strategy.

This chapter focuses on aspects of sustainable woody biomass production in *Quercus* dominated forest ecosystems with emphasis on different silvicultural management systems. Short rotation woody crops (SRWC), coppice with standards (CS), high forest (HF) and Satoyama are characterized according to their biomass potential and sustainability considerations. CS and HF are directly compared based on our own research and links to similar systems (SRWC and Satoyama) are drawn in order to provide a holistic view of the current topic. The chapter aims at providing an interdisciplinary view on biomass production in forest ecosystems, considering impacts on C and nutrient metabolism as well as other effects (e.g. biodiversity, technical, silvicultural and cultural issues). Considerations for sustainable biomass production in *Quercus* dominated forest ecosystems are presented for each management system.

1.1. Biomass production and carbon sequestration

Terrestrial C sequestration accounts for approximately one quarter of the three main sinks as indicated in Figure 2, where forests contribute the largest share. An intact terrestrial sink might be more important in the future in terms of mitigating climate change, since the ocean sink is expected to decrease. Our current forests are capable of sequestering ~2.4 Pg C yr^{-1} (of 2.6 Pg in total), when excluding tropical land-use change areas [10]. Sequestration of C in forests is controlled by environmental conditions, disturbance and management. However, forests can principally act as a source or sink of C, depending on the balance between photosynthesis and respiration, decomposition, forest fire and harvesting operations. On both European and global scales, forests were estimated to act as sinks on average over the last few decades [11, 12].

The most important process of net primary production is achieved by photosynthesis, which is the chemical transformation of atmospheric CO_2 and water from the soil matrix into more complex carbohydrates and long chain molecules to build up cellulose, which is found in cell walls of woody tissue as well as hemicellulose and lignin. The C remains in the woody compound either until it is degraded by microorganisms, which use the C as source of energy, or until oxidation takes place (e.g. burning biomass, forest fire). In both cases, it becomes part of atmospheric CO_2 again. A certain share, controlled mainly by climatic conditions [13], enters the soil pool as soil organic carbon (SOC). The ratio between aboveground and belowground pools depends on the current stand age, forest management and climate. In temperate managed forests, SOC stocks are typically similar to the aboveground stocks [14], which is confirmed in our own research [13]. SOC (and in particular O-horizon) stocks are typically higher in boreal forests and in high elevation coniferous forests as a consequence of reduced microbial activity and much lower in tropical environments. O-layer C pools are especially sensitive to changes in local climate. A traditional forest management regime in Austrian montane spruce forests is clear-cutting, typically from the top of a hill to the valley to facilitate cable skidding. An abrupt increase of radiative energy and water on the soil surface creates favourable conditions for soil microorganisms and a great amount of C stored in the O-layer will be released to the atmosphere by heterotrophic respiration. Other GHG's, such as N_2O are eventually emitted under moist and reductive conditions as excess nitrogen is removed by lateral water flows. Unfortunately, this effect is likely to happen on a large-scale where massive amounts of C might be released as the global temperature rises and thawing permafrost induces C emissions [15], potentially creating a strong feedback cycle, further accelerating global warming. However, Don et al. [14] found that SOC pools were surprisingly stable after a major disturbance (wind throw event), indicating low short-term vulnerability of forest floor and upper mineral horizons. They explained their findings with herbaceous vegetation and harvest residues, taking over the role of litter C input. The study covers a time-span of 3.5 years which might be too short to observe soil C changes. Likewise, we did not observe significant C stock decrease in our youngest sample plot of the coppice with standards (CS) chronosequence [13]. We expect that the N dynamics might have a profound influence on C retention and the impact of disturbance on SOC pools depends on environmental conditions. Successful long-term sequestration in terms of climate change mitigation is therefore only achieved if C becomes part of the recalcitrant fraction in the subsoil, which is typically between 1 000 and 10 000 years old [16, 17]. The C concentration is lower in the subsoil, but considerable amounts can still be found if one not only analyses the topsoil layer as recommended by a number of authors, e.g. by Diochon et al. [18]. In contrast, the radiocarbon age of the topsoil may range from less than a few decades [17] to months if considering freshly decomposed organic matter. The reasons for the relatively high age in subsoil horizons are not clear, but unfavourable conditions for soil microbial diversity and strong association of C with mineral surfaces (organo-mineral interactions with clay minerals) might be an explanation [16]. There are, in principal, two pathways for sequestering C in forest soils. Forest litter consists of leaves, needles and woody debris; such as branches, bark and fruit shells which accumulate on the surface (L and F layer). Soil macrofauna degrades it until it becomes part of the organic matter (OM) where it is impossible to recognize its original source (H layer). Parts of it are translocated into deeper horizons by bioturbation (e.g. earthworms) or remain on the surface, to be further degraded by soil microorganisms while organic matter becomes mobile in the form of humic acids and subsequently being mineralized at a range of negatively charged surfaces (humus, clay minerals). The second pathway is through root turnover and rhizodeposition (= excretion of root exudates). Matthews and Grogan [19] and subsequently Grogan and Matthews [20] parameterized their models with values of between 50 and 85% of C from the fine root pool which is lost to the SOC pool on an annual basis, depending on species composition and management. This assumption is consistent with another study where 50% of the living fine roots were assumed to reflect real values [21]. There is evidence that C derived from root biomass [22] and mycorrhizal hyphal turnover [23] might be the most important source for SOC pools rather than from litter decomposition. Since fine root turnover is species specific [24], it could be controlled to some extent by species composition at management unit levels. A further question for a forest owner in terms of carbon sequestration is which management option to choose while still being able to produce products and generate income. Despite the fact that unmanaged forests hold the highest C pools, it was commonly believed that aggrading forests reach a maximum sequestration and it is reduced in old growth forests, where photosynthesis and autotrophic respiration are close to offsetting each other. This is contrasts a number of studies, pointing out that even unmanaged forests at a late successional development stage could still act as significant carbon sinks [25]. The authors of another study claim that advanced forests should not be neglected from the carbon sequestration discussion a priori [26] and they should be left intact since they will lose much of their C when disturbed [27]. Therefore, managing forests implies a trade-off between maximum C sequestration and provision of goods, such as timber and biomass for energetic utilization. While the highest amounts of carbon could be sequestered in unmanaged forests [27, 28], C sequestration through forest management can be a cost-effective way to reduce atmospheric CO_2, despite limited quantities, due to biological limitations and societal constraints [29]. This is in general agreement with the conclusion of Wiseman's [30] dissertation, who argues that there is potential for additional C uptake depending on forest management, but the effect is short-term until a new equilibrium in C stocks is reached and also argues that the effect may be limited.

2. Biomass from forests

There is an on-going debate regarding the potentials of obtaining biomass from forests on multiple scales, from stand to international levels. Biomass is often discussed in the context of a raw material for energetic utilization although it should be emphasized that total biomass figures account for the total harvestable amount of wood, regardless of its utilization or economic value. Especially in the context of energy, it is highlighted that biomass is an entirely CO_2 neutral feedstock since the carbon stored in wood originates from the atmospheric CO_2 pool and it was taken up during plant growth. This is, in principal, true despite biomass from forests not being free of CO_2 emissions per se, since harvesting and further manipulation requires energy, which is currently provided by fossil fuels. However, it is difficult to estimate per-unit of CO_2 emissions since there are many influential variables. Even a single variable could have a profound influence on the per-unit emissions as is shown for the case of chipped fuel [31]. In general, biomass requires a different treatment as compared to fossil sources of hydrocarbons. Chemical transformations over thousands of years under high pressure led to a higher density of yieldable energy per volume unit as compared to biomass, although hydrocarbons are ultimately a form of solar energy. Hence fossil infrastructure does not fit to sources of renewable energy because of intrinsic properties. Centralized structures of energy distribution might work for fossil fuels, but it is questionable if it makes sense to transport woodchips across large distances. The energy invested for (fossil based) transport eventually curbs the benefits of renewable energy resources in terms of C emissions. Biomass from forests to be used for energetic utilization in the context of conventional forestry is often seen as a by-product of silvicultural interventions and subsequent industrial processes. However, there are a number of woodland management systems focussing on woody biomass production for energetic utilization or a combination of traditional forestry and energy wood production. Table 1 compares a number of *Quercus* dominated woodland management systems and highlights the main differences and characteristics. These systems will be further described thereafter.

In conventional forestry (high forest), residues from thinning and subsequent product cycles; e.g. slash and sawdust; are seen as the most important feedstock for energy wood. This opens the floor for controversial discussions and assumptions, based in principal on ecological and economic concerns. While residues of thinning operations are requested by traditional industries (e.g. paper mills), the extraction of slash and other harvest residues eventually leads to nutrient depletion with ecological impacts and ultimately detriment to increments in the long-term perspective. Inherent climate and soil properties control both magnitude and duration of such developments. "Residues" from forestry were traditionally harvested in ancient times. Most of the raw materials extracted from forests served as a source for thermal energy (fuel wood and charcoal) or other feedstock for industrial processes. Moreover, forests in central Europe provided nutrients for agro systems to sustain the human population [32]. Forest pasture, litter raking and lopping (sometimes referred to as pollarding) are some examples. Extraction of nutrients is still a common practice, e.g. litter collection in the Satoyama woodlands of Japan [33]. Since all of these practices tend to extract compartments with a relatively high nutrient content in comparison to wood, soil acidification and nutrient depletion was a common threat in Central European forest ecosystems. Forests only recovered gradually, mainly because of acidic depositions starting from the beginning of industrialization until the late 1980's, when clear signs of forest dieback caused public awareness and subsequent installation of exhaust filters across Europe.

Woodland management system	Characteristics
Coppice	<ul><li>Management target: traditional method of production of biomass for energetic utilization (fuelwood and charcoal)</li><li>Vegetative regeneration</li><li>Rotation period ~30 years</li><li>Degree of mechanization: low to medium</li><li>Extraction of nutrients from the soil: medium to high</li></ul>
Coppice with standards	<ul><li>Management target: coppice (underwood) e.g. for biomass and an uneven-aged structure of standards (overwood) for timber of high quality. Standards provide shade and act as a back-up if vegetative regeneration is not successful (seed trees).</li><li>Vegetative (coppice) and generative regeneration (standards)</li><li>Rotation period ~30 years for coppice and 60-120+ years for standards</li><li>Degree of mechanization: medium</li><li>Extraction of nutrients from the soil: medium to high</li></ul>
Short rotation woody crops (SRWC)	<ul><li>Management target: biomass for energetic utilization only, maximum increment will be harvested</li><li>Vegetative regeneration</li><li>Rotation period 2-10+ years (depending on the crop)</li><li>Degree of mechanization: high</li><li>Extraction of nutrients from the soil: very high</li></ul>
High forest	<ul><li>Management target: production of high quality stems for trade, biomass for thermal utilization is a by-product ("residues", thinning harvests, slash, sawdust)</li><li>Generative regeneration</li><li>Rotation period ~120 years</li><li>Degree of mechanization: medium</li><li>Extraction of nutrients from the soil: low to medium</li></ul>
Satoyama	<ul><li>Management target: integrative approach, biomass for energetic utilization, fertilizer for agriculture (litter collection) habitat for wildlife, recreation opportunities, small scale</li><li>Vegetative regeneration</li><li>Rotation period 15-20 years</li><li>Degree of mechanization: low</li><li>Extraction of nutrients from the soil: high to very high</li></ul>

Table 1. Comparison of *Quercus* woodland management systems and their characteristics with regard to management target, regeneration, rotation period, degree of mechanization and nutrient extraction. *Salix* or *Populus* species are the dominating crop in Central Europe due to higher increment rates in SRWC as compared to *Quercus* species.

Today, forest biomass stocks are increasing in most European countries, due to land use change (abandoned mountain pastures), shifting tree line as a consequence of global warming and elevated CO_2 concentrations as well as atmospheric N deposition. However, this should not lead to short sighted assumptions that biomass can be harvested at levels of growth increment, since a large part of it grows in areas with unsuitable conditions for access. Easily accessible forests at highly productive sites in lowlands are already typically managed at harvesting rates close to increment or even higher, e.g. in cases of natural disasters such as wind throws. In some countries, such as Austria, access to specific land ownership structures might uncover greater potentials of additional harvests.

3. Short rotation woody crops (SRWC) – A model of agriculture in forestry business

The major challenges of shifting forest management goals from traditional forestry to biomass production are sustainability issues, relatively low value of the product in comparison to quality logs and expensive harvesting, being competitive only at a high degree of mechanization in developed nations. As a consequence, rotation periods were shortened and fast growing species are preferred in order to produce woody biomass in an agriculture-like manner. Since the increment is highest at the beginning of stand development and subsequently decreases, only the maximum increment is utilized, ensuring maximum biomass production capacities at a given site. Short-rotation woody crops (SRWC) are hence established, in Europe typically with fast growing willow (*Salix*) or poplar (*Populus*) species. However, fast growing hardwood *Quercus* species are also considered for short rotation [34]. SRWC originated in ancient times, when people coppiced woodlands in order to obtain raw materials, e.g. fuel wood for cooking or heating purposes, but most of the research has been carried out and application of the results has been achieved in the last 50 years [35]. The basic principles of SRWC therefore originate in a coppice land management system, which will be described below. Planting is optimized for maximum biomass production (increment) while minimizing threats of disease and facilitating highly mechanized harvest technologies. Typical rotation periods are between 1 and 15 years [35], and rotations of *Salix* are shorter (< 5 years) than those of *Populus* and *Quercus*. Biomass from short rotations extracts significant amounts of soil nutrients, since a higher share of nutrient rich compartments (bark and thin branches) is extracted from the system. In combination with the short rotation cycles, nutrient extraction rates are larger as compared to conventional forestry. This implies the need of fertilization in most cases and concerns, e.g. about N leaching into groundwater bodies are discussed. However, Aronsson et al.[36] showed that high rates of N fertilization do not necessarily prime leaching, even on sandy soils (Eutric Arenosol) if the demand for N is high. C sequestration in the soil is also primarily controlled by N fertilization and the response of the vegetation [37]. The authors found increasing biomass production and C sequestration in a hybrid poplar plantation following N fertilization. On the other hand, it was argued that short rotations eventually result in the loss of the mineralization phase, thus preventing self-regeneration of the forest ecosystem [38]. Following their argument, a rotation cycle should be long enough to permit the return of autotrophic respiration and high rates of mineralization. In terms of C sequestration potential, it ultimately depends on the land use prior to SRWC if and to what extent additional C is accumulated. Especially sites that were formerly used for agricultural purposes and where organic carbon was depleted are prone to additional sequestration after land conversion [19, 20]. They pointed out that especially, but not only, non-woody *Miscanthus* plantations, can substantially sequester C with relative high amounts of litter. In a global context, SRWC may interfere with agricultural production if it continues to be a focus and if plantation areas increase since most plantations are not the result of forestland conversion, but rather farmland conversion. One of the reasons for this interference is the varying legal definition of SRWC across nations. While SWRC is considered forest in some countries, it is treated as an agricultural crop in other nations, making comparisons and predictions across borders difficult.

4. Coppice with standards and high forest management in Austria

High forest (HF) and Coppice with standards (CS) are the most common silvicultural management systems for broadleaved forest ecosystems in northeastern Austria. These systems have evolved over a long period of traditional management and they are mainly determined by environmental conditions and economic considerations. However, these systems were locally adapted over time, resulting in a range of intermediate types. Divergent silvicultural structures with diffuse standards are the consequence and are very common in Austria [39].

Quercus dominated high forests aim at producing quality timber with a diameter of at least 30 cm (diameter at breast height (DBH)). Rotation periods are approximately 120 years, followed by shelterwood cuts and natural regeneration. It is one of the most common systems in central Europe and suitable for most species. The rotation period is set according to the dominant or most economically significant species and it is usually shorter for coniferous species (~100 years). The most important difference in comparison to other management systems is the type of regeneration, which in the case of high forest is entirely generative. Generative regeneration may be introduced by shelterwood cuts or similar silvicultural systems, or by planting, while natural regeneration is usually preferred as a consequence of costs and genetic compatibility issues (e.g. climate) and uncertain provenance. Shelterwood cuts promote generative regeneration via seeds when the canopy is opened. Individuals with high quality (i.e. straight stem) may be chosen to initiate regeneration, which implies genetic selection to a certain extent. Thinning is typically performed 4 times, at 30, 50, 70 and 90 years. The main silvicultural goal is to produce straight logs with a minimum number and size of ingrown branches, as a raw material for woodwork, veneer and other similar purposes. Thinning operations and harvesting residues provide biomass for energetic utilization. Individual generative stems might be considered as a source of woody biomass for energetic utilization as well, if they do not meet requirements in terms of quality.

Quercus-Carpinus coppice with standards is a woodland management system to produce biomass for energetic utilization. These forests were once the main source of thermal energy when producing fuelwood for direct burning or charcoal production. There is evidence of coppice management dating back approximately 400 years in this region of Austria [40]. The management goals shifted during this period, depending on the demands of the landowners. CS is a relatively flexible system regarding supply of different qualities and quantities of wood. Among coppice, some trees are left in four age-classes to grow as larger size timber, called "standards". While standards provide a certain share of higher quality logs for trades, coppice provides fuelwood. Standards typically result from genetic regeneration. This multi-aged traditional system supports sustainable production of timber and non-timber forest products, while enhancing ecosystem diversity and wildlife habitat [41], which is also highlighted in the similar Japanese management system of Satoyama [33]. The rotation period for coppice (understorey) is typically 30 years [39, 42], hence holding a middle position between planted short rotation woody crops (SRWC) [35] and traditional high forests. The system is characterized by cyclic vegetative and generative regeneration [42]. Sprouting occurs rapidly after harvest and standards provide shade and are a source for seeds as backup if sprouting is not successful. Individual standards are managed in four age classes (30-60, 60-90, 90-120, and 120+ years) and harvested depending on certain criteria (e.g. market value, tree health, stand density). However, their importance began to cease with the introduction of fossil sources of energy during the onset industrialization, but significantly after 1960 [39]. Declining fuelwood demands led to a reduced intensity of understorey harvests (coppicing) and a shift towards longer rotation periods. This trend is especially distinct on fertile sites, while coppice was tendentially retained on sites with lower fertility.

The parent material of soils in our study region consists of gravel, sand and silt built up during the Pannonium (between 7.2 and 11.6 Ma before present) resulting from early formation of the Danube River. Consequently, a variety of soils can be found, e.g. Cambisols, Luvisols, Chernozems and even Stagnosols. Younger aeolian deposits of loess (Pleistocene) led to periglacial formation of Chernozems. The soils of our chronosequence series are classified as Eutric Cambisol with a considerable amount of coarse material (≤ 40% volume) in HF and sandy clay loam texture and both Haplic and Vermic Chernozems with loamy texture in CS [43]. Soils with lower fertility are derived from gravel and sand of the Danube River development, while Chernozems are derived from loess. The region receives approximately 500 mm of precipitation annually, with irregular periods of drought during summer. The water holding capacity of soils with a considerable amount of coarse material is lower as compared to loess derived soils, hence vegetative regeneration has the advantage of a fully functional root system at all times, supporting successful regeneration even in periods of drought. Generative regeneration might be obstructed under such conditions as a consequence of drying topsoil horizons. In our case study, we were able to include an outgrown coppice plot (i.e. a coppice with standards system that was not harvested at the theoretical end of the rotation period) aged 50 years to widen the scope for temporal dynamics. Irregular harvesting of standards and rotation periods up to 50 years (outgrown coppice) led to divergent silvicultural structures with diffuse standards [39], as previously mentioned. The plots were established during the summer of 2007 as permanent sample plots for aboveground biomass monitoring and are part of a framework to investigate biomass and carbon pools in this region [44].

4.1. Research methods

Studying temporal aspects of stand development is challenging because of the inherent duration of rotational cycles. Even in the case of CS in our example, it is theoretically 30 years but we included a 50-year-old outgrown stand. In HF, the rotational cycle is twice as long. Pickett [45] recommended a false chronosequence approach where he substitutes time for space. It is therefore crucial to find stands with similar management, species composition and other environmental conditions (microclimate, soil, topography), only differing in stand age. It must be assumed that the stands follow convergent succession trajectories [46], which was ensured in our case by use of inventory data from the past. The chronosequence approach is generally contested since it comes with a set of limitations (e.g. the problem of regional averaging, ignoring major disturbances or site-specific parameters as well as variation between hypothetical stands at the same age). Moreover, it assumes that there are no major disturbances (e.g. windthrows, insect attacks) during the rotational cycle. However, the method allows a researcher to successfully study temporal changes through the judicious use of chronosequences [46] and is often the only possible method to study long-term dynamics. Five plots were chosen for each chronosequence in HF and CS, ranging from 1-50 years in CS and 11-91 years in HF, respectively (Table 1).

A full biomass inventory was performed above- and belowground using allometric functions from Hochbichler [42]. In addition, belowground fine root biomass was determined to a depth of 50 cm by using soil cores. Additionally, soil macronutrient analysis was performed using these samples. Details for plot selection and setup as well as investigation of compartments and subsequent laboratory methods can be derived from Bruckman et al.[13]. The HF forest was chosen for comprehensive soil analysis, including exchangeable cations in soil and nutrient pools of different aboveground compartments, such as foliage, bark, wood and branches as well as regeneration and stems (see Figure 3). Exchangeable cations were determined at different soil horizons by using a $BaCl_2$ extraction and subsequent Inductively Coupled Plasma – Optical Emission Spectroscopy (ICP-OES) analysis. Exchangeable phosphorus pools were estimated using data from the forest soil inventory of adjacent forest sites [47]. Macronutrients N, P and K were determined in biomass compartments (foliage, bark of stems > 8 cm diameter, wood of stems > 8 cm diameter, composite sample of stems <8 cm, branches > 2 cm diameter, branches < 2 cm diameter and regeneration < 1.3 m height) according to inventory data for the most abundant species at each forest site. Nutrient analysis is based on three full tree samples (aboveground compartments) for *Quercus* and three foliage and branch samples from different crown layers per plot and species. Foliage was collected in August 2011 to ensure sampling fully developed leaves. $HNO_3/HClO_4$ extraction according to ÖN L 1085 [48] followed by ICP-OES analysis was used to determine nutrient contents.

4.2. Aboveground biomass

Aboveground biomass pools increased with stand age in both forest management systems as a consequence of steady accumulation (Figure 4). HF follows a typical pattern with high increments in the aggregation phase and marginal accumulation rates after 50 years.

Thinning concentrates additional growth on selected individuals with the highest possible quality, while low quality stems are harvested and utilized as fuelwood or further chipped. Hence the silvicultural activities aim at refining the product instead of maximizing biomass production. Approximately 2/3 of the aboveground biomass corresponds to understorey in the youngest HF site (11 years), while its share decreases steadily with increasing stand age (Table 1). The slightly rising share in the oldest stand (from ~3 to ~5 %) could be explained by initiation of generative regeneration as the canopy opens after thinning operations, allowing seeds to germinate and initiate regeneration.

Site	Age [years]	Biomass stocks [t/ha^{-1} dry mass]			Biomass stocks [%]	
		Overstorey	Understorey	Sum	Overstorey	Understorey
HF1	11	9.8	4.9	14.7	33.3	66.7
HF2	32	81	18.1	99.1	81.7	18.3
HF3	50	111.9	13.9	125.8	88.9	11.1
HF4	74	137.4	4.1	141.5	97.1	2.9
HF5	91	130.3	7.1	137.4	94.8	5.2
CS1	1	133.2	0	133.2	100	0
CS2	15	73.8	31	104.8	70.4	29.6
CS3	26	147.7	36.7	184.4	80.1	19.9
CS4	31	138.7	65	203.7	68.1	31.9
CS5	50	167.5	87.5	255	65.7	34.3

Table 2. Aboveground biomass stocks in tons per ha^{-1} dry mass for HF and CS, separated into overstorey and understorey compartments. In HF, overstorey represents individuals with DBH > 8 cm, while understorey represents individuals with DBH< 8 cm respectively. In CS, overstorey represents standards and understorey the vegetative coppice regeneration with some individuals being the result of generative regeneration.

In CS, we found a steady increase until the end of the rotation as the stand is still in the aggradation phase. The 15 year old stand is an exception because standards were previously harvested (irregular cut) resulting in lower biomass stocks as compared to the one year old stand where the total biomass equals that of standards. The relationship between overstorey and understorey biomass stocks is typical for coppice with standards forest management. While the overstorey stocks remain relatively constant (between 133 and 168 t.ha^{-1}), except in the 11 year old CS site where standards were recently harvested, the understorey coppice biomass pool constantly increases to 88 t.ha^{-1} at an age of 50 years (see Table 1). The relative share of coppice biomass increases from 20% in a 26-year-old stand to 34% in the oldest stand. This is an example for adaptive forest management since the demand for fuelwood has been low for decades and we were consequently able to find outgrown CS plots (50 years) and the share of coppice biomass is still relatively low. It could be increased under a different demand structure, where biomass for energetic utilization is in demand and commercialization becomes an interesting option for the forest owner.

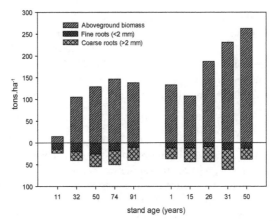

Figure 2. Aboveground (shoot) and belowground (root) biomass (dry mass) in two different forest management systems (high forest and coppice with standards). Data from Bruckman et al. [13].

On average, approximately 40% less biomass is stored in the HF system aboveground, and 7% less belowground (roots). Net primary production (NPP) is higher in CS which compensates lower basal area of the overstorey (DBH > 8 cm) with higher stand density [13]. The main reason of elevated NPP in CS is the higher fertility of chernozems as compared to cambisols in combination with a more effective water holding capacity as compared to the Eutric Cambisol in HF as a consequence of the coarse material content. The underlying silvicultural practices contribute to this biomass pool structure as thinning is performed at regular intervals in HF while typically only one intervention takes place in CS when harvesting coppice and selected standards at the end of the rotational cycle. As a consequence, additional C sequestration may only be achieved in CS when extending the rotation period. The second argument leads to higher productivity of the CS system, which allows higher C sequestration rates.

4.3. Belowground biomass

As previously mentioned, fine root turnover might be the most important pathway of C sequestration in forest ecosystems. Therefore, it is crucial to study root dynamics and turnover in the context of forest management. Our case study confirms that on average, fine root biomass (FRB) decreased with increasing stand age in HF (R = -0.28; p<0.01) but remained constant in CS. This basically reflects aboveground biomass dynamics where the CS system has a relatively balanced stand structure throughout rotational cycles. It is partly a consequence of shorter rotation cycles and therefore retains the aggradation phase [49]. Another reason lays in the continuous growing stock, since standards are kept on site during and after understorey harvest. Considering a finer resolution one may observe dynamic changes in FRB corresponding to stand development stages. FRB increased after stand reorganization, culminated at an age of 31 (CS) and 50 years (HF) and subsequently decreased as stands aged. In accordance with increasing aboveground biomass stores, coarse root C pools increased with age in HF (R= 0.87; p= 0.53), accounting for 8.0 (0.9) % of total C pool and no trend was observed in CS, where coarse root C pools accounted for 7.8 (1.0) % respectively [13]. Although on average HF has lower total belowground biomass stores (7 % less), the FRB is 32% higher as compared to CS. The root-to-shoot ratio indicates higher belowground relative to aboveground biomass accumulation rates in early successional phases. A direct comparison between HF and CS reveals two major differences:

1.   In comparison with HF, there was no initial major decrease of the ratio observed in CS
2.   The ratio is always lower in CS than in HF

These differences may be due to significant aboveground biomass stocks represented by standards in CS and therefore comparatively low ratios, even in the stand reorganization phase. Consequently, root/shoot ratios are in equilibrium throughout the rotation period. More favourable soil conditions in CS may lead to lower ratios throughout stand development. It was shown that drought and limited soil resources (nutrient) availability promote FRB production [50, 51]. The effect of standards harvesting was observed in our case study as a slightly higher ratio in the 15-year old stand compared with the one-year old stand in CS. On average, the root C pool represented 28.0(3.0)% of total phytomass C stores when excluding the youngest HF stand where the root C pool was 1.6 times as high as aboveground phytomass stores [13].

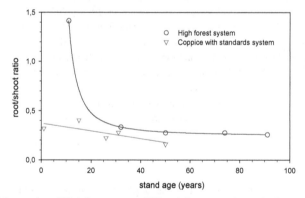

Figure 3. Root/shoot ratios of High forest system (HF) and Coppice with standards system (CS). The solid lines represent a hypothetic pattern. Source: Bruckman et al. [13].

4.4. Nutrient analysis

A comprehensive elemental analysis was performed for the HF system in both soil horizons and biomass compartments in the context of the framework for biomass investigations in northeastern Austria. HF was chosen because of lower soil fertility in comparison to the CS sites, which implies a higher sensitivity with regard to nutrient extraction.

The nutrient balance at a given site is an essential factor for stand productivity, species composition and biodiversity. Gains of nutrients that are in plant available forms and therefore could be incorporated in new plant tissue are limited to originate from weathering of bedrock material, atmospheric deposition as well as fertilization. The major processes of decreasing nutrient availability to plants are removal (biomass extraction and leaching) or chemical transformation processes, resulting in recalcitrant and thus plant unavailable fractions. Soil microbes play an important role in these processes and there is even a competition for some elements, e.g. N between microbes and plant roots [52]. Interfering nutrient cycles, e.g. by changing forest management practices therefore influences a complex system. The consequences are not usually recognized immediately, depending on the nutritional status of the soil. If nutrient pools and cation exchange capacity (CEC) are low, the consequences are seen within just a few years, as is the case in tropical soils, for instance.

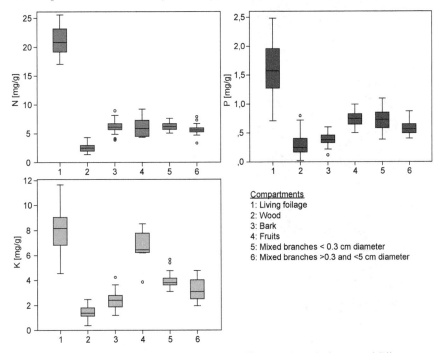

Figure 4. Nitrogen (N), Phosphorous (P) and Potassium (K) content in mg/g dry mass of different compartments for the dominating species, *Quercus petraea* in the HF system. Note that foliage is the compartment with the highest content of all macronutrients. Boxplots show the median (solid horizontal line), the bottom and the top of the box represent the 25th and the 75th percentile (interquartile range) and the whiskers show the highest and lowest values that are not outliers (< 1.5 times the interquartile range).

Based on an analysis of nutrient contents in various compartments (Figure 5), the aim was to compare plant available (exchangeable) pools of nutrient elements in soil with aboveground pools. This was done in order to determine potential nutritional bottlenecks when the management goal shifts towards biomass production as a source for energy, which implies higher nutrient extraction rates as compared to conventional forestry or intermediate types (see Table 1). We focussed on the dominant tree species (*Quercus petraea*, *Carpinus betulus* and *Corylus avellana*) where we sampled whole trees on each plot to account for local differences. Only foliage and branches were sampled from less abundant species (e.g. *Fagus sylvatica*, *Betula pendula* and *Prunus avium*) where they occurred.

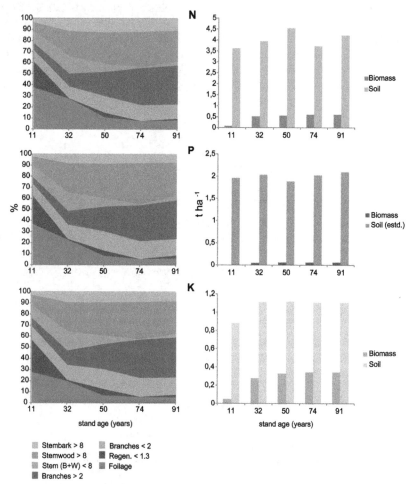

Figure 5. Relative amount of macronutrients in different compartments of *Quercus petraea* along the HF chronosequence (left) and a comparison of macronutrient contents in aboveground biomass and exchangeable soil pool (right). Category explanation (same order as legend): Bark of stems > 8 cm DBH, stemwood excluding bark of stems > 8 cm DBH, stems < 8 cm DBH (wood and bark), branches > 2 cm diameter (wood and bark), branches < 2 cm diameter (wood and bark), regeneration < 1.3 m height (total value), foliage (living). Soil exchangeable P pools were not actually measured, but estimated from data of the Austrian forest soil inventory [47].

Figure 5 illustrates the macronutrients nitrogen, phosphorous and potassium (NPK) contents of different compartments. Foliage clearly has the highest contents of macronutrients, followed by bark and thin branches. In the 91-year-old stand, foliage only accounts for 1.7% of the aboveground biomass, but represents 8.2% of the N pool, while 23.3% represent 37.5% in the youngest stand respectively. Wood (sapwood and heartwood) had the lowest contents. Similar patterns of nutrient distribution were previously reported for the same species [53]. The nutrient content of composite samples (wood and bark) depends on the respective proportions of wood and bark. However, it seems different for the case of P where higher contents in the composite as compared to separate wood and bark samples indicate higher contents of P in bark of thin branches (Figure 5). The bark sample consists of bark from branches and stem where the latter is therefore expected to have lower P contents. Approximately 40% of the macronutrients are stored in stems > 8 cm in diameter from an age of 50 years onwards (Figure 6) while representing approximately 60% of the stand aboveground biomass. Bark accounts for another 10% of the 40% stem pools. Consequently it was suggested to consider oak stem debarking to limit nutrient exports (especially Ca in the case of *Quercus* bark) from the stand [53]. A comparison with exchangeable soil pools revealed sufficient potential supplies from the soil matrix as the soil pools of macronutrients are well above the stand biomass pools. However, a simple comparison of pools does not necessarily represent the nutritional status of the vegetation since plant availability, stress and soil biogeochemical processes may cause uptake limitations of certain nutrients. For instance, the N:P ratio is well acknowledged as an indicator for either N or P limitation and values of < 14 indicate N deficiency where values of > 16 designate P limitation [54]. Obviously pools of soil exchangeable P are very high (Figure 6) which is also represented in our foliar N:P ratios. They are very stable at 14.2 for the 50, 74 and 91 year old stands and close to the threshold value (13.9) in the 32 year old stand. Interestingly the youngest stand (11 years) shows signs of N limitation with a ratio of 10.3. We suggest a combination of reasons for this observation. High rates of increment have a great N demand in the stand organization and subsequent aggradation phase. Herbaceous vegetation on this specific site might compete with woody species for topsoil N and the relatively high coarse material content contributes to low water holding capacities. In combination with low rates of precipitation, water availability might inhibit mobilisation and uptake of N at this site as suggested in a comprehensive review [34]. It is also shown that although the forest is surrounded by intensively managed agricultural land with associated atmospheric deposition of aerosols (dust), it has not led to P limitation as recently suggested [54]. In summary, despite evidence for N limits in the youngest stand, the macronutrient supply meets the demand under current forest management. However, it could be problematic if lower diameter compartments are extracted as biomass for energetic utilization. For instance, if the crown biomass is utilized at the final harvest and branches with a diameter > 2 cm are being extracted, it will account for a twofold extraction of N as compared to stem-only harvests. In a scenario of whole tree harvesting, close to 90% of N in biomass will be extracted, compared to approximately 40% in the stem-only scenario. Stem debarking could further reduce N extraction to below 30% of the total aboveground biomass pool. This example demonstrates the importance of assessing the dynamic nutrient pools in order to provide profound recommendations for sustainable forest management, as was concluded in a previous study [53]. Forest nutrition not only has implications on species diversity, sequestration of carbon, and provision for a magnitude of environmental services, but it has distinct implications on site productivity and thus earning capacity of a given management unit. Sustainable nutrient management is therefore an essential component of successful forest management, especially if management aims at harvests that are more intensive for bioenergy production.

5. Sustainable coppice biomass production: the Japanese example of Satoyama

Satoyama forests have a long tradition in Japan. Directly translated "sato" means "village" and "yama" "mountain" [55]. The translation points at the conceptual meaning of Satoyama, which describes the typical landscape between villages and mountains (Okuyama). Although there are many definitions, one probably finds a suitable one in the Daijirin dictionary: "the woods close to the village which was a source of such resources as fuelwood and edible wild plants, and with which people traditionally had a high level of interaction" [56]. Satoyama could be understood as an integrative approach of landscape management, including the provision of raw materials such as wood, natural fertilizer (see the transfer from nutrients from forests to agricultural systems as previously mentioned), drinking water and recreational opportunities. Besides its inherent economic and ecological values, it provides a sphere for human-nature interactions and as such, it opens a window to see how Japanese perceive and value their natural environment over time [56]. As a consequence of small-scale structures and specific management, Satoyama woodlands represent hotspots of biodiversity [55]. They are pegged into a landscape of paddy fields, streams and villages. From a silvicultural point of view, Satoyama woodlands consist of mainly deciduous species, such as *Quercus acutissima* and *Quercus serrata* (however, Japanese cedar is sometimes used to delimit parcels of different ownership), which are coppiced on rotational cycles of 15-20 years, creating a mosaic of age classes [33]. Along with their scenic beauty and recreational capabilities which is doubtlessly a strong asset close to urban megacities, Satoyama woodlands are capable of providing goods and services for the society; even under conditions of changing demands and specific needs of generations [57]. In contrast to Satoyama, coppice forests in Austria were traditionally managed to obtain fuelwood and because of suitable environmental conditions for coppicing. Low levels of precipitation (~500 mm yr^{-1}) promote coppicing, since a fully established root system is present at any time, stimulating re-sprouting even under periods of drought. However, the holistic approach of landscape management is by far not considered to the same extent as it is for Satoyama. The importance of coppice forests for biomass provision ceased with the utilization of fossil fuels. Likewise, Satoyama was devaluated as a result of the "fuel revolution" in Japan [57], during which large areas were abandoned, leading to dismissed or poorly managed Satoyama woodlands. This is comparable with the previously mentioned divergent silvicultural structures with diffuse standards in Austria [39]. Satoyama woodlands are now back in the public interest, starting in the 1980s when volunteer groups formed, being part of the Satoyama movement which urged for development of adequate environmental policies [33]. Although most groups are focussed on recreational activities in urban and peri-urban areas, the potential of Satoyama to provide biomass for energetic utilization has recently gained attention. Terada et al. [33] calculated a C reduction potential of 1.77 t ha^{-1} yr^{-1} of C for the Satoyama woodlands if coppiced and the biomass is utilized in CHP power plants. Considering the total study area (including settlements, agricultural areas and infrastructure, the C reduction potential is still 0.24 t ha^{-1} yr^{-1}. In the context of nutrient budgets, practices such as cyclic litter removal have to be evaluated with regard to the base cation removal rates and subsequent acidification, which ultimately leads to lower ecosystem productivity as was previously shown. Litter removal may cause substantial nutrient extractions of the compartment with the highest contents (Figure 5). The Fukushima Daiichi nuclear incident clearly showed the threat of nuclear fission energy sources. In combination with efforts to cut carbon emissions, paired with rising demands for energy, the situation caused a shift towards sustainable, clean and safe forms of energy provision. If a sustainable management is applied, especially in the context of nutrient budgets, Satoyama perfectly fulfils these requirements and might be a good choice to include in the future energy system while being neutral in terms of GHG release. Satoyama could represent a good model for sustainable resource management that the rest of the world can learn from [55, 56]. A study to evaluate Satoyama landscapes on a global basis was started under the "Satoyama initiative", launched by the Japanese Ministry of the Environment[*].

6. Conclusion

Specific types of biomass, i.e. wood and wood-derived fuels, have a long history of being the major source of thermal energy since humanity learned to control fire, which was a turning point in human development. These sources have not lost their significance in many developing countries, especially in domestic settings. However, over-population, climatic conditions and low efficiency cause shortages of fuelwood in many regions, e.g. Ethiopia or Northern India. This would not be the major problem in developed nations, where biomass as a source for thermal energy and raw materials for industrial processes has recently gained increased attention as a renewable and greenhouse-friendly commodity. Hence, sustainable management is required to prevent adverse consequences for society and the environment. Paradoxically biomass is tagged to be sustainable per se, although this is by no means substantiated since it depends on the local conditions and management used. Compared to conventional fossil sources of energy where "sustainability" is only directed at a wise and efficient use of a finite resource, sustainability of biomass from forests has to be considered in a much wider context. Biomass represents just one of a multitude of other ecosystem services and the potential for its provision depends on the conditions of any given ecosystem. As a matter of fact, ecosystems are extremely heterogeneous from global to forest stand scale, mainly controlled by environmental conditions, such as climate, soils and resulting species composition and anthropogenic impacts. Likewise, society's demands for specific ecosystem services are highly diverse. While recreation and the provision of a clean

---

[*] See http://satoyama-initiative.org/en/ for more details.

environment (water, air) are likely the most important service close to urban areas and settlements, the provision of wood and by-products as economic commodities might be important in more remote, but accessible regions. This also implies one of the major differences to the current energy system. Besides the fact that bio-energy is not capable of providing the same amount of sustainable energy we are currently receiving from fossil fuels, the infrastructure has to be decentralized, directed at local demands and supplies. Both, the economic stability as well as the benefits for environment and society are at risk in the case of large-scale bioenergy power plants. Based on a global review, Lattimore et al. [58] identified six main areas of environmental constraints with regard to wood fuel production:

1. Soils
2. Hydrology and water quality
3. Site productivity
4. Forest biodiversity
5. Greenhouse gas balances
6. Global and supply-chain impacts of bioenergy production

This listing elucidates the challenge of applying sustainability criteria to biomass production, since a large set of criteria and indicators for bioenergy production systems has to be implied. Consequently, sustainability comes at the cost of a high complexity of ensuring mechanisms. Nonetheless, a set of regionally adaptable principles, criteria, indicators and verifiers of sustainable forest management, as suggested by Lattimore et al. [58] might be inevitable to ensure best practices according to the current status of scientific knowledge. They propose an adaptive forest management framework with continuous monitoring in certification systems to ensure efficacy and continual improvement. Policy has to ensure that such regulations are implemented in binding regulations. Corruption and greed for short-term profit maximization are still major problems, providing a base for unsustainable use of resources, especially in countries with unstable political situations. Apart from the general challenges of biomass production, we showed a number of examples of different *Quercus* dominated forest management systems and their considerations for sustainable biomass production.

Short rotation woody crops (SRWC) have the potential to maximize biomass production, which comes at certain environmental costs. Utilizing the maximum possible increment, by using fast growing species including *Quercus* and by reducing rotation periods, soils have to provide substantial amounts of nutrients, which must be returned by fertilization. Economic considerations (economies of scale) and high levels of mechanization lead to uniform structures and monocultures. Hence, biodiversity and a number of other ecosystem services are threatened. Fertilization in agriculture is always associated with considerations of groundwater pollution. Likewise, fertilization can have the same effects in SRWC if soil properties are ignored and improper fertilization strategies are applied. However, there are indications that SRWC may better utilize the nutrients from fertilization and leaching could be minimized [36]. This might be a consequence of perennial crops in comparison to conventional crops in agriculture and deeper rooting of the biomass crops. However, another potential problem linked to agriculture is the fact that SRWC are often established on former agricultural land. Together with the trend towards other kinds of agrarian non-woody biomass crops for energy production (biofuels, biogas), it should be noted that agricultural land should be given priority for food and feed production under current predictions for population growth. This conversion of use is especially problematic in some developing countries and was recently discussed extensively [59].

Quercus dominated high forest (HF) represents a system of conventional forestry and it was expedited as a consequence of reduced demands for fuelwood and charcoal as they were substituted by fossil sources of energy. Another reason was increasing demands for higher timber qualities for construction and trade. This reversal in forest management towards longer rotation periods of > 100 years in combination with limited utilization of small diameter compartments such as branches and twigs resulted in improving soil conditions in many cases in Europe, especially close to urban areas. However, the potential for biomass production for energetic utilization is limited as this system aims at biomass refinement rather than maximizing outputs in terms of quantities. Potential sources for bioenergy production are harvests of silvicultural interventions (thinning) as well as crown and stump biomass at the end of the rotation period. We showed that crown biomass removal in addition to stem utilization (full tree harvest) will extract 50% more N as compared to stem-only harvesting. On the contrary, stem debarking could retain 70% of the total aboveground N pool in the system under a scenario of stem-only extraction. In accordance with aboveground stand development, the root-to-shoot ratio follows a distinct pattern. While the woody vegetation invests more in root growth in the youngest stand to explore soil nutrient and water reserves, the ratio reaches a value below 1 (i.e. aboveground biomass prevails) in the 32 year old HF stand and subsequently levels off. This observation was supported by the N:P ratio in the youngest HF indicating N limitations, potentially priming root growth. In general, the biomass potential for energetic utilization is lower as compared to the CS system, which is mainly expressed by higher NPP as a consequence of more fertile soils and different management aims. If more intensive biomass extractions are considered to be sustainable or not therefore depends on soil nutrient and water budgets and has to be assessed at a local scale. Our HF sites showed sufficient nutrient pools despite some indication for N deficiency in the youngest plot.

Coppice with standards (CS) holds an intermediate position between SRWC and HF. It is capable of providing both higher quality timber logs and biomass for energetic utilization. The system has a long tradition in our study region and it is very flexible concerning different demands of the respective qualities and quantities. It was demonstrated that CS has the advantage of a fully functional root system at all times during the rotational cycles which is represented in the relatively stable root-to-shoot ratio and this is an advantage in relatively dry climates. Re-sprouting occurs relatively fast and is likely to be more successful as compared to planting or natural generative regeneration, especially under conditions of seasonal droughts. However, standards also act as seed trees where genetic selection is possible (promotion of individuals with high stem quality). They act as a backup if vegetative regeneration is unsuccessful and provide shade during summer. Nutritional bottlenecks are not expected under current management practices as the soils in our CS plots are relatively fertile and the forestland is surrounded by extensively managed agricultural land. However, if the management strategy is changed towards schemes of increased biomass extraction, the effects on soils have to be studied in order to ensure sustainability.

The Satoyama woodlands in Japan are an example of traditional sustainable woodland management, carefully balancing ecosystem services in regions with relatively high population density, thus implying a high level of social-nature interaction. Since this form of landscape management proved to be successful over centuries with a high degree of flexibility with regard to changing demands over time, it might be a model of sustainable land use for other regions in the world. However, the effects of cyclic litter removal from soil nutrient pools should be investigated since we demonstrated that foliage is the compartment holding the highest contents of macronutrients.

A sustainable future, entirely based on renewable sources of energy without harming our environment is possible. It will certainly base on decentralized structures with a large pool of different sources of renewable energy, which has another great advantage of a high level of resilience in comparison to our current system. Biomass can play a significant role in areas with sufficient supplies, as long as the production follows sustainability criteria and does not interfere with other environmental services essential for that region. The optimal future energy system consists of a range of different sources, in which biomass is eligible along with other renewable sources as long as it is produced in a sustainable manner. Certainly, changing lifestyles (reduced energy consumption, less meat in diets, higher efficiency) especially in the developed regions of the world may be a very important and effective step to reducing resource consumption, which should be taken immediately.

Author details

Viktor J. Bruckman* and Gerhard Glatzel
Austrian Academy of Sciences (ÖAW),
Commission for Interdisciplinary Ecological Studies (KIOES), Vienna, Austria

Shuai Yan
Northwest Agricultural and Forestry University, College of Forestry, Yangling, China

Eduard Hochbichler
University of Natural Resources and Life Sciences (BOKU), Institute of Silviculture, Vienna, Austria

Acknowledgement

This study was funded by the Austrian Federal Ministry of Agriculture, Forestry, Environment and Water Management (Project No. 100185) and the Austrian Academy of Sciences (ÖAW), Commission for Interdisciplinary Ecological Studies (KIÖS) (Project No.

---

* Corresponding Author

P2007-07). We would like to thank Robert Jakl and Stephan Brabec for their unremitting support in fieldwork and lab analysis as well as Yoseph Delelegn and Arnold Bruckman for their invaluable efforts in collecting samples in the field. Toru Terada enriched our study with valuable discussions on coppice management in Japan. We would further like to express our gratitude to Christina Delaney for proofreading of the manuscript.

7. References

[1] Le Quere C, Raupach MR, Canadell JG, Marland G, et al. Trends in the sources and sinks of carbon dioxide. Nature Geosci. [10.1038/ngeo689]. 2009;2(12):831-6.

[2] International Energy Agency. World Energy Outlook 2011 - Executive Summary. Paris: OECD Publishing; 2011.

[3] Kiehl JT, Trenberth KE. Earth's Annual Global Mean Energy Budget. Bulletin of the American Meteorological Society. 1997 1997/02/01;78(2):197-208.

[4] Peters G, Marland G, Le Quéré C, Boden T, Canadell P, Raupach M. An update of the global carbon budget: Emissions rebound after the Global Financial Crisis (GFC). Planet under Pressure 2012; London, UK, 2012.

[5] Tans P. Trends in Atmospheric Carbon Dioxide. National Oceanic and Atmospheric Administration (NOAA), Earth System Research Laboratory (ESRL), USA; 2012.

[6] Neftel A, Moor E, Oeschger H, Stauffer B. Evidence from polar ice cores for the increase in atmospheric CO_2 in the past two centuries. Nature. [10.1038/315045a0]. 1985;315(6014):45-7.

[7] Plass GN. The Carbon Dioxide Theory of Climatic Change. Tellus. 1956;8(2):140-54.

[8] Canadell JG, Le Quéré C, Raupach MR, Field CB, Buitenhuis ET, Ciais P, et al. Contributions to accelerating atmospheric CO_2 growth from economic activity, carbon intensity, and efficiency of natural sinks. P Natl Acad Sci USA. 2007 November 20, 2007;104(47):18866-70.

[9] Lambin EF, Meyfroidt P. Global land use change, economic globalization, and the looming land scarcity. P Natl Acad Sci USA. 2011 Mar 1;108(9):3465-72.

[10] Pan Y, Birdsey RA, Fang J, Houghton R, Kauppi PE, Kurz WA, et al. A Large and Persistent Carbon Sink in the World's Forests. Science. 2011 August 19, 2011;333(6045):988-93.

[11] Nabuurs GJ, Thurig E, Heidema N, Armolaitis K, Biber P, Cienciala E, et al. Hotspots of the European forests carbon cycle. Forest Ecology and Management. 2008 Jul 30;256(3):194-200.

[12] Vetter M, Wirth C, Bottcher H, Churkina G, Schulze ED, Wutzler T, et al. Partitioning direct and indirect human-induced effects on carbon sequestration of managed coniferous forests using model simulations and forest inventories. Global Change Biology. 2005 May;11(5):810-27.

[13] Bruckman VJ, Yan S, Hochbichler E, Glatzel G. Carbon pools and temporal dynamics along a rotation period in *Quercus* dominated high forest and coppice with standards stands. Forest Ecology and Management. 2011;262(9):1853-62.

[14] Don A, Bärwolff M, Kalbitz K, Andruschkewitsch R, Jungkunst HF, Schulze E-D. No rapid soil carbon loss after a windthrow event in the High Tatra. Forest Ecology and Management. 2012;276(0):239-46.

[15] Schuur EAG, Abbott B. Climate change: High risk of permafrost thaw. Nature. [10.1038/480032a]. 2011;480(7375):32-3.

[16] Schmidt MWI, Torn MS, Abiven S, Dittmar T, Guggenberger G, Janssens IA, et al. Persistence of soil organic matter as an ecosystem property. Nature. 2011 Oct 6;478(7367):49-56.

[17] Rumpel C, Kogel-Knabner I, Bruhn F. Vertical distribution, age, and chemical composition of organic, carbon in two forest soils of different pedogenesis. Org Geochem. 2002;33(10):1131-42.

[18] Diochon A, Kellman L, Beltrami H. Looking deeper: An investigation of soil carbon losses following harvesting from a managed northeastern red spruce (Picea rubens Sarg.) forest chronosequence. Forest Ecology and Management. 2009 Jan 31;257(2):413-20.

[19] Matthews R, Grogan P. Potential C-sequestration rates under short-rotation coppiced willow and Miscanthus biomass crops: a modelling study. Aspects of Applied Biology. 2001;65:303-312.

[20] Grogan P, Matthews R. A modelling analysis of the potential for soil carbon sequestration under short rotation coppice willow bioenergy plantations. Soil Use and Management. 2002 Sep;18(3):175-83.

[21] Richter DD, Markewitz D, Trumbore SE, Wells CG. Rapid accumulation and turnover of soil carbon in a re-establishing forest. Nature. 1999;400:56-8.

[22] Rasse D, Rumpel C, Dignac M-F. Is soil carbon mostly root carbon? Mechanisms for a specific stabilisation. Plant and Soil. 2005;269(1):341-56.

[23] Godbold D, Hoosbeek M, Lukac M, Cotrufo M, Janssens I, Ceulemans R, et al. Mycorrhizal Hyphal Turnover as a Dominant Process for Carbon Input into Soil Organic Matter. Plant and Soil. 2006;281(1):15-24.

[24] Jha P, Prasad Mohapatra K. Leaf litterfall, fine root production and turnover in four major tree species of the semi-arid region of India. Plant and Soil. 2010;326(1):481-91.

[25] Schulze E-D, Wirth C, Heimann M. Managing Forests After Kyoto. Science. 2000 September 22, 2000;289(5487):2058-9.

[26] Knohl A, Schulze ED, Kolle O, Buchmann N. Large carbon uptake by an unmanaged 250-year-old deciduous forest in Central Germany. Agricultural and Forest Meteorology. 2003 Sep 30;118(3-4):151-67.

[27] Luyssaert S, Schulze ED, Borner A, Knohl A, Hessenmoller D, Law BE, et al. Old-growth forests as global carbon sinks. Nature. [10.1038/nature07276]. 2008;455(7210):213-5.

[28] Schulp CJE, Nabulars GJ, Verburg PH, de Waal RW. Effect of tree species on carbon stocks in forest floor and mineral soil and implications for soil carbon inventories. Forest Ecology and Management. 2008 Jul 30;256(3):482-90.

[29] Seidl R, Rammer W, Jager D, Currie WS, Lexer MJ. Assessing trade-offs between carbon sequestration and timber production within a framework of multi-purpose forestry in Austria. Forest Ecology and Management. 2007 Aug 30;248(1-2):64-79.

[30] Wiseman CLS. Organic Carbon Sequestration in Soils: An Investigation of Five Profiles in Hesse, Germany [Dissertation]. Frankfurt am Main: Johann Wolfgang Goethe - Universität; 2003.

[31] Van Belle JF. A model to estimate fossil CO_2 emissions during the harvesting of forest residues for energy-with an application on the case of chipping. Biomass & Bioenergy. 2006 Dec;30(12):1067-75.

[32] Glatzel G. The impact of historic land use and modern forestry on nutrient relations of Central European forest ecosystems. Nutr Cycl Agroecosys. 1991;27(1):1-8.

[33] Terada T, Yokohari M, Bolthouse J, Tanaka N. Refueling Satoyama Woodland Restoration in Japan: Enhancing Restoration Practice and Experiences through Woodfuel Utilization. Nature and Culture. 2010;5(3):251-76.

[34] Salon C, Avice J-C, Alain O, Prudent M, Voisin A-S. Plant N Fluxes and Modulation by Nitrogen, Heat and Water Stresses: A Review Based on Comparison of Legumes and Non Legume Plants. In: Shanker AK, Wenkateswarlu B, editors. Abiotic Stress in Plants - Mechanisms and Adaptations: Intech Science; 2011.

[35] Dickmann DI. Silviculture and biology of short-rotation woody crops in temperate regions: Then and now. Biomass & Bioenergy. 2006 Aug-Sep;30(8-9):696-705.

[36] Aronsson PG, Bergstrom LF, Elowson SNE. Long-term influence of intensively cultured short-rotation Willow Coppice on nitrogen concentrations in groundwater. Journal of Environmental Management. 2000 Feb;58(2):135-45.

[37] Garten J, C.T., Wullschleger SD, Classen AT. Review and model-based analysis of factors influencing soil carbon sequestration under hybrid poplar. Biomass & Bioenergy. 2011;35:214-26.

[38] Trap J, Bureau F, Vinceslas-Akpa M, Chevalier R, Aubert M. Changes in soil N mineralization and nitrification pathways along a mixed forest chronosequence. Forest Ecology and Management. 2009;258:1284-92.

[39] Hochbichler E. Methods of oak silviculture in Austria. Annals of Forest Science. 1993;50(6):583-91.

[40] Frank J. Der Hochleithenwald, Einführung in die Wälderschau des Niederösterreichischen Forstvereins 1937 in das Rudolf Graf von Abensberg und Traunsche Forstrevier Wolkersdorf: Selbstverlag des NÖ. Forstvereins. 1937.

[41] Nyland RD. Silviculture: concepts and applications. 2. ed. New York: McGraw-Hill; 2002.

[42] Hochbichler E. Fallstudien zur Struktur, Produktion und Bewirtschaftung von Mittelwäldern im Osten Österreichs (Weinviertel). Wien: Österr. Ges. für Waldökosystemforschung und Experimentelle Baumforschung Univ. für Bodenkultur; 2008.

[43] IUSS Working Group WRB. World reference bases for soil resources 2006. Rome: FAO2006.

[44] Hochbichler E, Bruckman VJ, Spinka S, Glatzel G, Grieshofer H. Untersuchungen zur Dynamik der Biomassen- und Kohlenstoffvorräte in Niederwäldern mit Überhälter, Mittel- und Hochwäldern. Vienna: University of Natural Resources and Life Sciences, Department for forest and soil sciences 2009. Report No.: 100185.

[45] Pickett STA. Space-for-time substitutions as an alternative to long-term studies. In: Likens GE, editor. Long-term Studies in Ecology. New York: Springer; 1989. p. 110-35.

[46] Walker LR, Wardle DA, Bardgett RD, Clarkson BD. The use of chronosequences in studies of ecological succession and soil development. J Ecol. 2010 Jul;98(4):725-36.

[47] Englisch M. Österreichische Waldboden-Zustandsinventur. Teil III: Atmogene Hauptnährstoffe. Österreichische Waldboden-Zustandsinventur Ergebnisse. Vienna: Forstliche Bundesversuchsanstalt; 1992. 247 pages.

[48] Chemical analyses of soils - Extraction of elements with aqua regia or nitric acid - perchloric acid mixture, ÖNORM L 1085 (1999).

[49] Bormann FH, Likens GE. Pattern and process in a forested ecosystem disturbance, development and the steady state based on the Hubbard Brook ecosystem study. New York: Springer; 1979.

[50] Santantonio D, Hermann RK. Standing crop, production, and turnover of fine roots on dry, moderate, and wet sites of mature Douglas-fir in western Oregon. Annals of Forest Science. 1985;42(2):113-42.

[51] Noguchi K, Konopka B, Satomura T, Kaneko S, Takahashi M. Biomass and production of fine roots in Japanese forests. J Forest Res-Jpn. 2007 Apr;12(2):83-95.

[52] Gärdenäs AI, Ågren GI, Bird JA, Clarholm M, Hallin S, Ineson P, et al. Knowledge gaps in soil carbon and nitrogen interactions – From molecular to global scale. Soil Biology and Biochemistry. 2011;43(4):702-17.

[53] Andre F, Ponette Q. Comparison of biomass and nutrient content between oak (*Quercus petraea*) and hornbeam (*Carpinus betulus*) trees in a coppice-with-standards stand in Chimay (Belgium). Annals of Forest Science. 2003 Sep;60(6):489-502.

[54] Heilman P, Norby RJ. Nutrient cycling and fertility management in temperate short rotation forest systems. Biomass and Bioenergy. 1998;14(4):361-70.

[55] Morimoto Y. What is Satoyama? Points for discussion on its future direction. Landscape and Ecological Engineering. 2011;7(2):163-71.

[56] Knight C. The Discourse of "Encultured Nature" in Japan: The Concept of Satoyama and its Role in 21st-Century Nature Conservation. Asian Studies Review. 2010 2010/12/01;34(4):421-41.

[57] Yokohari M, Bolthouse J. Keep it alive, don't freeze it: a conceptual perspective on the conservation of continuously evolving "Satoyama" landscapes. Landscape and Ecological Engineering. 2011;7(2):207-16.

[58] Lattimore B, Smith CT, Titus BD, Stupak I, Egnell G. Environmental factors in woodfuel production: Opportunities, risks, and criteria and indicators for sustainable practices. Biomass and Bioenergy. 2009;33(10):1321-42.

[59] Winiwarter V, Gerzabek MH, editors. The Challenge of sustaining soils: Natural and social ramifications of biomass production in a changing world. Vienna: Austrian Academy of Sciences Press; 2012.

Chapter 6

Harvest Systems and Analysis for Herbaceous Biomass

Jude Liu, Robert Grisso and John Cundiff

Additional information is available at the end of the chapter

1. Introduction

Biomass is a distributed energy resource. It must be collected from production fields and accumulated at storage locations. Previous studies of herbaceous biomass as a feedstock for a bioenergy industry have found that the costs of harvesting feedstocks are a key cost component in the total logistics chain beginning with a crop on the field and ending with a stream of size-reduced material entering the biorefinery. Harvest of herbaceous biomass is seasonal and the window of harvest is limited. Biomass needs to be stored at a central location. Normally, several or many of these central storage locations in a certain range of a biorefinery are needed to ensure 24 hours a day and seven days a week supply. These centralized storage locations are commonly called satellite storage locations (SSL). The size and number of SSLs depend on the size of the biorefinery plant, availability of biomass within a given radius, window of harvest, and costs.

It is convenient to envision the entire biomass logistics chain from fields to biorefineries with three sections. The first section is identified as the "farmgate operations", which include crop production, harvest, delivery to a storage location, and possible preprocessing at the storage location. This section will be administrated by farm clientele with the potential for custom harvest contracts. The second section is the "highway hauling operations," and it envisions commercial hauling to transport the biomass from the SSL to the biorefinery in a cost effective manner. The third section is the "receiving facility operations," and it includes management of the feedstock at the biorefinery, control of inventory, and control of the commercial hauler contract holders to insure a uniform delivery of biomass for year-round operation.

Agricultural biomass has low bulk density, and it is normally densified in-field with balers, or chopped with a self-propelled forage harvester. Currently, there are four prominent harvesting technologies available for biomass harvesting [1]. They are: (1) round baling, (2)

rectangular baling, (3) chopping with a forage harvester, and separate in-field hauling, and (4) a machine that chops and loads itself for in-field hauling (combined operation). Large round and large rectangular balers are two well-known and widely accessible harvesting technologies [2], which offer a range of advantages and disadvantages to farmgate operations. Round bales have the ability to shed water. When these bales are stored in ambient storage, they will store satisfactorily without covering and storage cost is significantly reduced. The round baler, because it is a smaller machine with fewer trafficability issues, can be used for more productive workdays during an extended harvest season over the winter months [3]. Large rectangular bales have greater bulk density, ease of transport, and increased baler productivity (Mg/h). However, the increased capital cost for the large rectangular baler and the bales' inability to shed water limit its use on farms in Southeastern United States. Bale compression machines are available to compress a large rectangular bale and produce high-density packages [4]. The densified package has two or three times higher density than the field density of large rectangular bales [5].

The goal of an effective logistics system is to streamline storage, handling, and preserve the quality of the biomass through the entire logistics chain. This goal will minimize average feedstock cost across year-round operation. The farmer shares the goal to preserve the quality of the biomass, and also desires to produce the biomass at minimum cost. To assist in the accomplishment of the mutual objectives of both parties, this chapter will discuss major logistics and machine systems issues starting from the farmgate to the receiving facilities at a biorefinery. Constraints in this biomass supply chain will also be discussed. The impact of different harvest scenarios for herbaceous biomass harvest will be shown. Logistics systems have been designed for many agricultural and forest products industries. Thus, it is wise to use the lessons learned in these commercial examples. Each of these industries faces a given set of constraints (length of harvest season, density of feedstock production within a given radius, bulk density of raw material, various storage options, quality changes during storage, etc.), and the logistics system was designed accordingly. Typically, none of these systems can be adopted in its entirety for a bioenergy plant at a specific location, but the key principles in their design are directly applicable. Commercial examples will be used in this chapter to interpret these principles.

2. Biomass harvesting and the field performance of harvest machine systems

2.1. Introduction to herbaceous biomass harvesting

Harvesting of cellulosic biomass, specifically herbaceous biomass, is done with a machine, or more typically a set of machines, that travel over the field and collect the biomass. These machines are designed with the traction required for off-road operation, thus they typically are not well suited for highway operation. Therefore, the transition point between "in-field hauling" and "highway hauling" is critical in the logistics system. In-field hauling is defined as the operations required to haul biomass from the point a load is created in-field to a storage location chosen to provide needed access for highway trucks. This hauling includes hauling in-field plus some limited travel over a public road to the storage location.

Harvesting systems can be categorized as coupled systems and uncoupled systems. Ideal coupled systems have a continuous flow of material from the field to the plant. An example is the wood harvest in the Southeast of the United States. Wood is harvested year-round and delivered directly to the processing plant. Uncoupled systems have various storage features in the logistics system.

Sugarcane harvesting is an example of a coupled system for herbaceous crops. The sugar cane harvester cuts the cane into billets about 38-cm long and conveys this material into a trailer traveling beside the harvester (Figure 1). The harvester has no on-board storage. Thus, a trailer has to be in place for it to continue to harvest. The trailer, when full, travels to a transfer point where it empties into a truck for highway hauling (Figure 2). Each operation is coupled to the operation upstream and downstream. It requires four tractors, trailers, and operators to keep one harvester operating. The trucks have to cycle on a tight schedule to keep the trailers moving. One breakdown delays the whole operation.

A "silage system" can be used to harvest high moisture herbaceous crops for bioenergy. With this system, a forage harvester chops the biomass into pieces about one inch (25.4 mm) in length and blows it into a wagon beside the harvester. This wagon delivers directly to a silo (storage location), if the field is close to the silo, or it dumps into a truck for a longer haul to the silo. All operations are coupled. That is, a wagon must be in place to keep the harvester moving, and a truck must be in place at the edge of the field to keep the wagons cycling back to the chopper. It is a challenge to keep all these operations coordinated.

A coupled system can work very efficiently when an industry is integrated like the sugarcane industry in South Florida, USA. Because the sugar mill owns the production fields surrounding the mill and the roads through these fields, the mill controls all operations (harvesting, hauling, and processing). Sugarcane has to be processed within 24 hours after harvest so the need for a tightly-controlled process is obvious.

Figure 1. Sugar cane harvester delivering material into a dump trailer for delivery to edge of field (Photo by Sam Cooper, courtesy of Sugar Journal, P.O. Box 19084, New Orleans, LA 70179).

Figure 2. Transfer of sugar cane from in-field hauling trailers to highway-hauling trucks.

An example of an uncoupled system is cotton production using the cotton harvester that bales cotton into 7.5-ft diameter by 8-ft long round bales of seed cotton. This system was developed to solve a limitation of the module system. With the module system, in-field hauling trailers (boll buggies), have to cycle continuously between the harvester and the module builder at the edge of the field. The best organized system can typically keep the harvester processing cotton only about 70% of the total field time. Harvesting time is lost when the harvester waits for a trailer to be positioned beside the harvester so the bin on the harvester can be dumped.

Baling is an uncoupled harvest system and this offers a significant advantage. Harvesting does not have to wait for in-field hauling. Round bales, which protect themselves from rain penetration, can be hauled the next day or the next week. Rectangular bales have to be hauled before they are rained on.

2.2. Field capacity and efficiency of biomass harvest machines

The equipment used for baling and in-field hauling is a critical issue to the farm owners. More efficient harvest systems coupled with well-matched harvesting technologies specific to farm size and crop yield can minimize costs. The importance of understanding the linkage between various unit operations in the logistics chain was illustrated [6,7]. Similarly, researchers have quantified the handling and storage costs for large square bales at a bioenergy plant [8]. However, in both of these evaluations the details of field operations and field capacities of machines involved in the field harvesting and handling were not available [9]. Instead, costs of bales at the farm gate were used to analyze bioenergy production costs. To maximize the field efficiency of field machine systems, it is essential for farm managers to know the field capacity of each machine involved in harvesting. In addition, quantitatively understanding the capacity of biomass harvest machines is essential to assess daily production and supply rate for a biorefinery or a storage facility.

The field efficiency of rectangular balers can be determined by calculating the theoretical material capacity of the baler and actual field capacity [10]. Calculation of material capacity can be demonstrated using a large rectangular baler as an example. The end dimensions of the large bales were 1.20 m × 0.90 m; bale length was 2.44 m. The depth and width of the chamber were 0.9 m and 1.2 m, respectively. Plunger speed was 42 strokes per minute. Measured bale density was 146 kg/m^3, and the thickness of each compressed slice in bale was 0.07 m. Thus, calculated theoretical bale capacity using equation 11.60 in [10] is 27.83 Mg/h. The plunger load could be set higher and produce higher density bales.

The theoretical capacity was obtained from a baler manufacturer under ideal conditions. Ideal conditions exist when a baling operation has [11]:

1. Long straight windrows
2. Windrows prepared with consistent and recommended density (mass/length)
3. Properly adjusted and functioning baler
4. Experienced operator

Actual field capacity of a baler will be impacted by the size and shape of the field, crop type, yield and moisture content of the crop at harvest, and windrow preparation. Typical field efficiencies and travel speeds can be found from ASAE Standards D497 [12]. Cundiff et al. [11] analyzed the field baler capacity and considered the effect of field size on baler field capacity. They found that the field capacities of round and large rectangular balers were 8.5 Mg/h and 14.4 Mg/h, respectively.

Another example of testing the baling capacity of a large rectangular would be the field tests conducted on wheat straw and switchgrass fields [13,14]. Results showed that actual field capacity of a large rectangular baler was between 11 and 13 Mg/h. This indicates that the field capacity of a large rectangular baler could be 50% or less compared to its theoretical capacity.

2.3. Power performance of harvest machine systems

Machine system field efficiency is limited by tractor power performance, machine field capacity, and field conditions. Field conditions limit operational parameters and the percentage of the maximum available power. Since the high cost of harvest and in-field handling is still one of the main roadblocks of utilizing biomass feedstocks to produce biofuels, increasing machine system field efficiency through designing or selecting suitable machine systems is the challenge to machinery design and management professionals.

Power performance of a 2WD rear drive tractor was presented in a format of flow chart in ASAE standards [12,15]. Total power required for a tractor is the sum of PTO power (P_{pto}), drawbar power (P_{db}), hydraulic power (P_{hyd}), and electric power (P_{el}) as expressed by equation (1). Depending on the type of implement, components in equation (1) may vary. Total power calculated with equation (2) is defined as equivalent PTO power, which can be used to estimate the tractor fuel consumption under specific field operations.

$$P_T = \frac{P_{db}}{E_m E_t} + P_{pto} + P_{hyd} + P_{el} \qquad (1)$$

Where E_m and E_t are mechanical efficiency of the transmission and tractive efficiency, respectively. Each of the power requirements in Equation (1) can be estimated using recommended equations in [12]. Example of using this standard to estimate power requirements for a large rectangular baler system is available [13].

2.4. System analysis case study - Large rectangular bale handling systems

2.4.1. Baling and bale collection

Baling, bale collecting, and bale storing of 650-ha wheat straw field at a commercial farm was studied as a typical large rectangular bale harvesting system. Factors that affect large rectangular bale production and handling logistics were quantified through observing complete field operations and then the system performance was analyzed and field capacities of all machines in this system was determined. System limitations were quantified, and means to reduce production costs were discussed [14].

Bales were stored in covered storage facilities. The equipment used included one large rectangular baler, two bale handlers, and three flatbed bale trucks (Figure 3). The straw was raked with a twin rotary rake. The rake was used to form uniform and evenly-spaced windrows from the straw that had been expelled by the combines. If the straw was rained on before baling, the rake fluffed the straw to enhance drying. The operator adjusted the swath of the rake in order to form windrows of proper size.

Figure 3. The large square baler, bale handler, and bale truck used in straw harvesting.

The field crew included a person operating the baler, two persons operating wheel loaders, and three truck drivers. The baler ran continuously throughout the day with the exception of operator breaks. Since no bales were left in the field overnight, the bales were collected at about the same pace as produced. A truck driver would bring the truck into the field and locate two bales. The wheel loader operator followed the truck through the field and loaded two bales at a time. When the truck was full, the driver would exit the field and transport the load to a storage facility. Bales were loaded in an interlocking manner, and the bales were not strapped down, a procedure that saved a lot of time.

Because there were three truck drivers and two trucks, a driver was not present while the truck was unloaded. The driver of the fully loaded truck would drive into the storage facility and position the truck to be unloaded. By the time this truck arrived with a full load of straw, the previous truck at the site would be empty. This process was repeated, with minor delays when a driver waited for a truck to be unloaded. Capacity of each operation was measured in terms of number of large rectangular bales (Table 1). Baling rate was the limiting factor in the system.

2.4.2. Bale compression

To reduce long distance hauling cost of forage/hay bales, compression of these bales into ultra-dense bales was adopted by some producers to prepare the hay for overseas transport.

Commercial bale compression equipment has potential to be used for the densification of baled biomass.

Field operation	Number of bales per hour (St. dev.)
Raking	85 (equivalent)
Baling	43 (9)
Accumulating into stacks of bales in field	93 (28)
Loading truck in field from accumulated bales	204 (29)
Loading truck in field from not first accumulated bales	143 (46)
Stacking bales in storage	93 (12)

Table 1. Field capacities of large rectangular baler and bale handlers (Straw M.C., 14% w.b.) [14].

Ten large rectangular switchgrass bales were compressed using the compressor (Figure 4) [9]. The input bale had a dimension of 0.86×1.22×2.21m, and the resulting compressed bales were 0.53×0.46×0.38m. When stacked in groups of 20 per pallet, the package had a dimension of 1.02×1.17×1.70m. The initial bales had a volume of 23.27m^3. When accounting for a 7.0% loss of material and water, this volume was 21.67m^3. The bales had average moisture of 14% (w.b.) at the time of compression. The compressed package had a volume of 13.13m^3, a reduction of 60.6% in volume.

Figure 4. Commercial large rectangular bale slicer/compressor.

2.5. System analysis case study – Round bale handling system

The advantage of round bales is that the rounded top sheds water and bales can be stored in ambient storage without the expense of covered storage. A second important advantage of round bales is that round balers are conventional technology throughout the United States. They are widely used to harvest forage. Compared to a large rectangular baler, a round baler has lower capital cost.

A significant advantage is realized by the farmers if their round balers can be used for both their existing livestock enterprise and a bioenergy enterprise. Warm-season grasses are harvested during the winter for the bioenergy market. Thus, the biomass harvest does not conflict with the hay harvest. The advantage gained by the biorefinery is that no need for their feedstock producers to invest in new equipment. Requiring capitalization of new equipment will make the contacting with feedstock producers more difficult.

The disadvantage of round bales is that they are hard to stack and thus the handling and transport costs are higher. System performance was analyzed and field capacities were determined based on field measurements [16]. Individual handling of bales (either round or rectangular) is not cost effective. The high cost is caused by long loading and unloading times. Multi-bale handling units have been designed, and these units are discussed in Section 9 of this chapter.

3. Field working days – Harvesting limitations

Determining the schedule time required for harvest operations is an essential prerequisite to determine herbaceous biomass priority for securing inventory and effectively matching the transport of biomass from SSLs (satellite storage locations) to a biorefinery. Estimating the number of days expected to be available for baling is difficult because agricultural field work is heavily weather dependent and information concerning winter field operations is not well established. Harvest costs depend in part on the investment required in harvest machines (number of machines purchased), and this investment depends on the field capacity of the machines and the number of field workdays available during the harvest window. Therefore, an estimate of the number of harvest days is necessary to determine the total investment in harvest machines required to support a biorefinery.

The optimal perennial grass harvest strategy for a biorefinery remains to be determined. One potential strategy is to delay harvest until the plants standing in the field have transitioned from a non-dormant to a dormant state. Transitional events within the plant include translocation of nutrients from the above-ground plant parts to below-ground plant parts as initiation of degeneration (senescence) of above-ground tissues occur. Delaying harvest until the end of this transitional period, and harvesting over a short time period, can result in near maximum biomass yield, reduced moisture content, and reduced amount of nutrients removed with the biomass. This strategy would suggest an optimum harvest window for switchgrass from October to December. However, a harvest season this short would require substantial investment in harvest machines to complete harvest within a 3-month period. It also requires a large storage for inventory to supply the plant for year-around operation. By delaying harvest until nutrients have translocated, biomass tonnage will be decreased by this loss of mineral content, however since the relative amount of carbon in the material is increased, conversion efficiency and combustion quality may be improved [17]. Moreover, the translocated nutrients stored in the roots can be used for growth and development by the plant year after year, thus, reducing the need for and cost of supplementing the soil with nutrients through fertilization.

A second strategy is to extend harvest over as many months as possible. This would enable more economical use of harvest machines, reduce the quantity required in SSL storage, and potentially better match truck transport with SSL operations. With an extended harvest strategy, harvest would begin in July and extend through the following March. This extended harvest season would allow for spreading the fixed costs of the harvest machines over more Mg harvested per year thus reducing the $/Mg harvest cost. However, the earlier harvests before senescence (July through September) will remove more nutrients, and they will potentially conflict with other farm operations (i.e. grain/hay harvest, small grain seeding). To maintain productivity, additional fertilizer will be required, and this represents an additional cost to the farmer whose fields are harvested prior to October. The farmer whose fields are harvested January through March will have a lower yield because of leaf loss from the standing biomass exposed to winter conditions.

A delayed harvest raises an important issue relative to the payment to a farmgate contract holder. As previously explained, a certain loss occurs when the crop is left standing in the field for delayed harvest (December through March). Also, the delayed harvest increases the risk that a weather event will cause the crop to lodge so the harvest machine leaves material, and the "yield" is reduced. If these same fields were harvested during the optimum window (October through December) and stored longer in an SSL, there is a storage loss that increases with time in storage [18]. Considering the total reliance on stored biomass during the non-harvest months, which strategy is the better choice for the farmgate contract holder (i.e., generates the maximum quantity of biomass to sell to the biorefinery)? Is it better to delay harvest or build a larger SSL and invest in more harvesting machine capacity to complete the entire harvest within the 3-month optimum window? Epplin et al. [19] estimated differences in switchgrass harvest costs for both a 4-month and a 8-month harvest window. They accounted for differences in biomass yield across months, but they did not adjust for fertilizer requirements. They found that the estimated harvest costs varied from $25 per Mg for a 4-month harvest season to $11 per Mg for a 8-month harvest season. These results show that the length of the harvest season is a significant factor in determining the costs of harvesting biomass.

Epplin et al. [19] did not have refined estimates of the number of days that biomass could be harvested. They based their estimates of available harvest days on a study designed to determine the number of days that farmers in Southwestern Oklahoma of United States could conduct tillage operations. To determine more precise estimates of the number of harvest machines required to harvest, and thus obtain a more precise estimate of harvest costs, a more precise estimate of the number of harvest days is required.

Hwang et al. [20] determined the number of suitable field workdays in which switchgrass could be mowed and the number of days that mowed material can be baled. Empirical distributions of the days available for mowing and for baling switchgrass were determined for nine counties in Oklahoma. Distributions were determined for each month and for two potential harvest seasons (short, October–December and extended, July - February). Several conditions are necessary for safe baling of switchgrass. First, the soil must be sufficiently dry to support the weight of harvest machines. Second, prior to baling, the moisture content of the cut switchgrass must be at a level for safe storage. They provided evidence of the difference in harvest days across strategies.

Determining the time available for required harvest operations is a necessary, but not sufficient, prerequisite for determining the optimal harvest strategy. Additional information will also be required. As previously explained, if harvest begins in July prior to maturity and subsequent senescence in October, additional fertilizer will be required (the amounts are undefined) to offset the nutrients removed from the pre-senescence harvest. If harvest is delayed into December or later, the harvestable yield will be less (the amounts are undefined) than if it is harvested in October and November. Thus, the information regarding differences in harvest days across months must be combined with information regarding differences in fertilizer requirements, and biomass yields across harvest months, to determine an optimal harvest strategy.

4. Constraints for a typical biomass logistics system

4.1. Constraints set by resource

Any biomass crop that provides an extended harvest season has an advantage because most biorefineries want to operate as many hours per year as possible. Continuous operation yields the maximum product per unit of capital investment. Herbaceous crops cannot be harvested during the growing season. Thus, storage is always a component in a single-feedstock logistics system design.

Two examples are discussed from opposite ends of the length-of-harvest-season spectrum. In the Southeastern United States, wood is harvested year-round. This is referred to as "stored on the stump." Weather conditions in the Southeast are such that few harvest days are lost in the winter due to ice and snow. On the other end of the spectrum, consider the harvest of corn stover in the Upper Midwest of the United States. The grain harvest is completed in the fall, and then the stover is collected. In some years, there are less than 15 days between the completion of the grain harvest and the time when the fields are covered with snow, and no stover can be collected. All feedstock required for year-round operation of a bioenergy plant, if it uses only corn stover, must be harvested in a three- to five-week period. This is a significant challenge for a logistics system.

4.2. Constraints set by purchaser of raw biomass

Biorefinery operators are interested in the cost of the feedstock as it enters their plant. The plant can burn the biomass directly to produce heat and power, or it can use the biomass as a feedstock for some more complex conversion processes to produce a high-value product. The term "feedstock" is used to refer to any raw biomass before its chemical structure is modified by a conversion process, which can be direct combustion, thermo-chemical, or biological.

Feedstock cost ($/dry ton) is defined as the cost of the stream of size-reduced material entering the reactor at the Bioenergy plant for 24/7 operation. The reactor is defined as the unit operation where an initial chemical change in the feedstock occurs. The reason for choosing this reference point for the biomass logistics system is two-fold.

1. The plants that can operate continuously, 24/7, have an advantage. Maximum production (tons/y) per unit of capital investment gives a competitive advantage in the market place – cost to produce the product is lower.

2.   Some logistics systems do size reduction with the harvesting machine, while some size reduce at a transfer point between in-field hauling machines and highway hauling machines (perhaps as a pre-requisite to a densification step), and some size reduce at the entrance to the processing plant. In order to compare the several systems, it is necessary to have a consistent end point for the system analysis.

The readers should be aware that many studies in the literature select a different analysis endpoint than used here. A typical endpoint is the cost of feedstock when a truckload of raw biomass enters the plant gate. This reference point is favorable as many agricultural and forest industries pay the producer when the raw material is delivered to the plant.

5. Field loss

The yield losses of different herbaceous biomass feedstocks while standing in the field will impact the harvest systems and the windows of opportunities to gather the feedstocks into storage. The inventory losses will be impacted by herbaceous biomass feedstocks, package configurations and integrity, storage facilities and time in storage. Loss of dry matter is also an important parameter in biomass collection and transportation.

Biomass loses dry matter due to its high moisture or dryness. Leaves and other fragile parts of the plant are broken and lost in the wind or mixed with soil. Some of the losses occur during storage due to fermentation and breakdown of carbohydrates to carbon dioxide and other volatiles. Unfortunately the exact account of switchgrass losses in the field or during storage is not available. Sanderson et al. [21] reported dry matter loss in baling and storage of switchgrass and stated that the overall losses were less than for legume hay. They estimated that switchgrass bales stored outside without protection resulted in a dry matter loss of 13% of the original bale dry weight. They also estimated that dry matter loss of 1-5% during baling depending on the moisture content. Kumar and Sokhansanj estimated field and storage losses for straw and stover [22]. Other studies have also estimated the dry matter loss of biomass during storage, collection and transport [23-26]. Turhollow [27] estimated the losses from switchgrass to be similar to losses in alfalfa. The study estimated 8% losses for a mower-conditioner, 3% for a rake, 10% for a round baler, and 0.1% for a round bale wagon. He estimated average loss of 15% over 6 months of storage. A recent study showed that the dry matter loss in switchgrass collection (including storage) is less than 2% for different collection methods.

Moisture causes damage (microorganism growth) and subsequent dry matter losses in stored switchgrass bales. Several studies have shown that dry matter losses in switchgrass bales are greater for bales stored outside as compared to bales stored inside [18,22,28]. Moreover, dry matter losses are far greater for covered rectangular bales than uncovered round bales [18]. Large uncovered round bales had a better economic return than covered rectangular bales, when considering the cost due to mass loss during storage [18]. However, another study highlighted their successful use of rectangular bales [29]; the cost of covered storage was more than offset by the reduction in hauling cost for the square bales.

6. Computer model of biomass logistics systems

When a series of operations are linked into a long chain of events, it is often difficult to see which operation is having the most impact on cost-effectiveness. Computer-assisted decision making tools are useful to select the most cost-effective logistics system. Recent studies [30-33] compared a number of different scenarios to determine the best unit operation options and identify bottlenecks. Mathematical programming approaches have been used to develop optimization models that are applicable to a variety of cases studies [34-37]. More recently, focus has shifted to simulation models using object-oriented programming [38] or discrete event simulation [39, 40]. The object-oriented approach [22, 41-43] simplifies scenario building, particularly related to various equipment options available for the same operation, since data pertaining to different options are stored in a standard object-oriented format.

The biomass logistics subject area can be divided into two types of modeling approaches depending on the environment encountered: stochastic or deterministic, and integer or continuous. In the areas of stochastic and deterministic environment, ISBAL and BioFeed capture the stochastic biomass logistics issues such as variability in processing as a result of weather, time, equipment breakdowns, etc. Additional work in this area can be found in [34; 40; 44-45]. In the deterministic environment, most work utilizes a geographic information system (GIS) interactively with optimization. This allows the GIS framework to work as a data management tool, which may call the optimization software directly. Most of the deterministic models are mixed-integer programs [37, 46-49]. One of the principles of designing an effective logistics system is starting with the feedstock resources distribution, and the simplest method is to utilize a GIS framework for data management.

Continuous models assume that all the land within a region is utilized for biomass production [50-52]. This approach uses average haul distances and cost. Hence, the solution obtained is difficult to implement due to lack of detail, as opposed to the solution obtained using an integer model formulation, where each production field is considered as a specific entity with specific costs, thereby resulting in precise decisions for implementation.

6.1. Modeling of biomass harvesting and handling systems for field operations

There are numerous studies developing for the optimum set of field equipment where the unit operations affect the performance of units upstream or downstream. Review of this literature is beyond the scope of this chapter. However, several simulation programs are highlighted that specifically addresses a biomass harvest and delivery system.

The Integrated Biomass Supply Analysis and Logistics Model (IBSAL) is a simulation model which simulates operations in the field [22,38,41]. ISBAL is very useful for the simulation of operations that collects feedstock from the field and examines the flow of biomass into storage. IBSAL is best used for drawing conclusions about a sequential set of events. For example, if the user has a defined number of fields, a defined set of equipment, and a target number of tons to be harvested each month (each week) then IBSAL can provide valuable guidance for optimum biomass collection. The influence of weather, equipment breakdowns, and other disturbances to the biomass system can be "played" with a series of simulations.

The BioFeed model [43] determines the overall system optimum, and integrates the important operations in the biomass chain into a single framework. It is possible that the optimal solution recommended by BioFeed may not be implemented in a real system, either due to unforeseen disturbances such as weather or due to the actions of independent stakeholders such as farmers. However, an integrated model such as Bio-Feed determines the optimal configuration, system bottlenecks, and potential improvements. Such a model can be useful in quantifying the systemic impacts of technology improvement [42].

The BioFeed model results were compared to recent studies in literature [22,31,33]. The scope of the BioFeed model was similar to the scenarios developed in [22], where the delivered cost was estimated to be about \$35/Mg (d.b.) excluding biomass size reduction. The major differences between ISBAL and BioFeed were harvesting and storage costs. While Kumar and Sokhansanj [22] ignored the storage costs, they also considered a self-propelled forage harvester which had a higher throughput capacity than the mower-conditioner considered in BioFeed, thereby reducing the cost. Khanna et al. [31] reported the switchgrass delivered cost of \$64.84/Mg, which was similar to the BioFeed cost estimate. The study by Duffy [33] estimated the delivered cost to be \$124.30/Mg. However, the major difference in the estimates was due to a much higher establishment cost.

Since the scope of the analyses and the assumptions differed for these studies, it is impossible to make specific comparisons. However, these comparisons illustrate that the overall model predictions agree reasonably well.

6.2. Modeling of biomass delivery systems

Linear programming models have been used to analyze system interactions in biomass delivery systems. Dunnett et al. [53] proposed a program to optimize scheduling of a biomass supply system for direct combustion. This model simulated storage on farms and delivery to one location with a variable demand for heat. They suggest that costs of biomass handling can be improved 5 to 25% with the model recommendations. Bruglieri and Liberti proposed a "branch and bound" nonlinear model to determine biorefinery locations as well as the optimum transport method [54]. Their model focused on multiple feedstocks but did not use actual equipment performance data.

A model comprised for multiple purposes can bring attributes of benchmarking, simulation, and linear programming together to solve for the best solution. Leduc et al. a system of wood gasification plants optimized [55]. Their model focused on establishing a biorefinery plant in a location that is suitable for distribution of the product being manufactured (in this case, methanol). Other similar models have focused on silage handling operations [56].

A number of models have proven that a single chain of handling procedures can be optimized, but fail to adequately address the "disconnect" caused by storage, specifically satellite storage. Unfortunately, few models consider different harvest systems (or feedstock) supplying a single biorefinery. For example, one equipment system delivers chopped material directly from the field to the plant (probably from fields close to the plant), and a second set of equipment bales material and places it in storage which will be delivered during months when direct delivery of field chopped material is not possible.

As described earlier, a satellite storage location (SSL) is a pre-designated location that is used as a storage location for the biomass collected by the producers within a defined geographic region. SSLs are a logical transition point between "agricultural" and "industrial" operations and thus are critical elements in a logistics system design.

6.3. Study of Satellite Storage Locations (SSLs)

An SSL should be established where sufficient feedstock density will ensure the investment is justified. Location affects the costs associated with transporting the biomass from the production field to the SSLs and from the SSLs to the biorefinery. An additional factor that needs to be considered is how often a SSL is filled and unloaded. If a SSL is emptied only once a year, then the total SSL storage area would be doubled as compared to a twice-per-year unload schedule. Note, multiple fillings of a SSL is possible if harvest can be extended over several months and is properly matched with the SSL unload sequence.

Specialized equipment, with high productivity (ton/h) will be utilized to empty each SSL. This equipment can be permanent equipment for each SSL, or the equipment may be mobile and move from one SSL to the next. Questions that need to be answered to implement the mobile option are: 1) how many sets of equipment should be used to service the entire area? 2) What is the sequence of SSLs that each equipment set should unload? 3) Is an SSL ready to be unloaded? This last question directly affects the hauling company's contract. These questions are best considered in a virtual environment where scenarios can be compared and contrasted.

A case study to characterize feedstock resources and establish the SSL's was analyzed by Resop et al. [57] for a 48-km radius around Gretna, VA, USA. The GIS analysis identified potential production fields based on current land utilization determined using aerial photography and landuse classification (Figure 5). The analysis selected fields such that the production area was 6% of the total land area within the radius. The biomass produced is sufficient to supply the demand for a 1,944 Mg/d biorefinery, assuming 47 operating weeks per year. SSLs were established at 199 locations, and the existing road network (GIS database) was used to determine the travel distance from each SSL to the proposed plant location in Gretna. A weighted Mg-km parameter for transportation from the SSLs to the plant was computed to be 44.8 km, which implies that, averaged across all 199 SSLs, each Mg traveled an average of 44.8 km to the biorefinery.

Judd et al. studied three equipment options for the operations performed at a SSL [58]. Two options utilize the rack concept [59]. Ravula et al. analyzed a round bale logistics system utilizing this rack system; the rack size emulate a 20-ft (6.1-m) ISO container providing two levels of 8 bales, total of 16 bales, to be handled as a single unit [40]. A tractor-trailer truck can haul two racks (32 bales, 14.4 Mg). The results showed the rack hauling had higher transportation costs due to not loading the truck to maximum allowable weight.

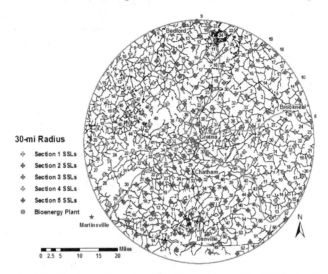

Figure 5. Example of Satellite Storage Locations (SSLs) located over a 30-mi (48-km) radius around a chosen bioenergy location. The refinery is located in the center of the circle. Each cross represents an SSL location with access to the public road system. The smallest SSL was a storage that can store biomass from 60 ac (24 ha) of production fields and the largest stores biomass from 1200 ac (486 ha) of production fields.

At a SSL, the round bales could be handled in one of two ways. The first option loads bales into the rack from the rear and is referred to as the "rear-loading." The second option, referred to as the "side-loading," loads the bales into a rack from the side. These two options were compared to grinding the bales at the SSL and compacting the ground material into a briquette (maximizes over-the-road load) for delivery [50]. Judd et al. [58] found that it was more cost effective to use the "side-load" option and haul the round bales than to form briquettes at the SSL, if the haul distance was less than 50 miles (81 km).

The large biorefinery concept calls for delivery of feedstock from a large geographic area. The results [58] suggest that some type of intermediate processing step, referred to as an Advanced Regional Biomass Processing Depot (ARBPD) by Eranki et al. [60], may be needed. These depots will convert the raw biomass to a more energy dense (higher value) product, and this product will then be delivered to a large biorefinery for final processing.

Why was the expense of size reduction and densification into briquettes at the SSL considered? Independent studies [61-63] reported that the tradeoff between the additional cost for densification and the reduction in transportation cost is significant for the hauling of logging residues. For additional work in the area of densification, see [64-65].

6.4. Application of information technologies

The inclusion of a hauling contractor in the business plan provides the best opportunity for all of the technology developed for other logistics systems to be applied to feedstock logistics. The information technologies applied include a GPS receiver in every truck and a bar code on every load as well as geographic mapping and routing management tools. Thus, time and location information can be observed at every SSL and the weigh-in scale at the receiving facility. The data collected can be used to optimize asset utilization in real time, and it will also feed needed data into the accounting software to compensate the farmgate and haul contracts.

It is expected that the collected data will be presented in real time. The information will provide a "Feedstock Manager," maps showing the location of all assets with updates at suitable time intervals. The goal is to provide an opportunity for the Feedstock Manager to make optimization decisions in real time. Examples are: trucks rerouted to avoid traffic delays, assets redeployed during breakdowns, increase at-plant inventory when inclement weather is forecasted, and a turn-down of plant consumption when a delay in feedstock deliveries cannot be avoided.

Some perspective of the logistics complexity, as presented to the Feedstock Manager, can be gained from the following sample parameters. This example presumes that operations will be in the Upper Southeast of the United States where switchgrass is harvested over an 8-month harvest season. Suppose the supply area has 199 SSLs within a 48-km radius of the plant, and each SSL has a different amount of material stored as shown in Figure 6. The producer desires to fill their SSLs at least twice during the harvest season to minimize per-Mg SSL investment. Suppose there are five SSL crews under contract and each of them wants the same opportunity to earn income (total Mg mass hauled per year). The Feedstock Manager's job is to treat all farmgate and haul contractors fairly.

6.5. Benchmarking models for biomass logistics

It is appropriate to benchmark a proposed feedstock delivery system against existing commercial systems; for example, woodchips, grain, hay, sugarcane, and cotton. Ravula et al. [37] studied delivery to a cotton gin from over 2,100 field locations to better understand logistics system design for a biorefinery.

Switchgrass, corn stover, and other energy crops present two major challenges when benchmarked against other commercial examples. First, energy crops are spread over a greater number of sites when compared with woodchips harvested as byproducts from logging. Second, energy crops have low value when compared with hay, grain, sugarcane, or cotton. Applying benchmarked models to energy crop systems provides a starting point, but optimization techniques must be applied to "fine tune" the logistics system design.

6.6. Simulation of mixed harvesting options for supply of single plant

Brownell and Liu [66] developed a computer model to select the best option, or the best combination of options, for supply of a given bioenergy plant. The options include a direct chop and delivery option, two bale options (round and large rectangular), and a variation of the large rectangular bale option where these bales are compressed at a stationary location before shipment. Specifically, the options are: (1) loose biomass harvest using a self-loading forage wagon; (2) round bale harvest; (3) large rectangular bale harvest; and (4) large rectangular bale with compression before hauling.

The results of the model provide users a decision matrix which shows the optimized handling scenario for all four handling options analyzed by this program. Cost calculation was based on three plant demands (2,000, 5,000, and 10,000 Mg DM/d) to compare costs as the required production area expanded. Satellite storage locations were established based on the distribution of production fields and the existing road network.

The key constraints used for the simulation are:

1. Only loose biomass can be directly hauled from the field to the biorefinery.
2. Biomass cannot be stored at the plant, 20,000 tons (18,144 Mg) will "stand in field" in surrounding area of the plant and collected on demand as loose material.
3. Baled biomass will be transported to and stored at satellite storage locations.
4. Large rectangular bales may be compressed before transportation.
5. Biomass will be grown on one third of the available acreage, with an average yield of 3 Mg/ha (15% average moisture content).

The land close to the plant is more valuable, from the biorefinery viewpoint, for the production of feedstock. The study found that a plant can afford to pay more for feedstock (landowner gets a higher price) if a farm happens to be closer to the chosen plant location.

The results showed that the size of SSLs is sensitive to the amount of material demanded by the plant. The program tests various sizes, distances between SSLs, and possible overlaps of production fields "served" by each SSL. The programming optimizes the simulation by slowly changing the size and location of SSLs. The program solves for the maximum amount of acreage available in the simulation model. The output is a distance material in each SSL can be efficiently hauled based on if it is harvested and hauled directly from the field or if it is baled and stored before hauling. The program also solves for the lowest cost for a set acreage when the closest fields are field chopped and remaining fields are baled.

7. Commercial examples of logistics systems

7.1. Typical linkage in logistics chain

As previously stated, biomass logistics is "a series of unit operations that begin with biomass standing in the field and ends with a stream of size-reduced material entering a bioenergy plant for 24/7 operation." This concept can be visualized as a chain (Figure 6), where the links are unit operations. The example in Figure 6 shows a logistics system moving bales from the field to SSLs, and then from these SSLs to a bioenergy plant. The dotted lines show various segments of the chain that are assigned to the several entities in the business plan. In this example, the segment identified "farmgate" is performed by the feedstock producer, the segment labeled "SSL" is performed by a load-haul contractor; the segment identified as "Receiving Facility" is performed by the bioenergy plant. Figure 6 is one example; several other options are used in commercial practice. In next section, three commercial examples are given to show readers the range of options in a logistics system.

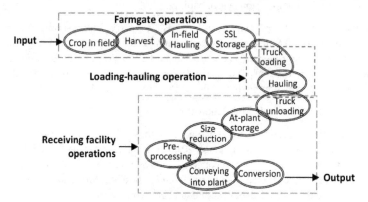

Figure 6. Logistics chain for delivery of round bales to a bioenergy plant (The dotted lines identify segments of the chain which are performed by different business entities).

7.2. Examples of commercial logistics models

7.2.1. Traditional model

The producer grows, harvests, stores, and delivers raw material (biomass) in accordance with a contract with the processing plant. Deliveries are made to insure the plant has a supply for continuous operation during the processing season. For many agricultural industries (i.e. cotton, grain, sugar cane, fruits and vegetables), the processing season approximately coincides with harvest season, and it is only part of the year.

The advantage of the traditional model, from a processor's point of view, is that all quality issues reside with the producer. There is no question who is responsible if a quality standards are not met. Business people planning the operation of a bioenergy plant tend to prefer this model, though they typically balk at paying a feedstock price that adequately compensates the producer for their additional risk.

7.2.2. Cotton model

A cotton producer grows, harvests, and stores the raw material (seed cotton) in modules at an edge-of-field location with suitable access for highway hauling trucks (Figure 7). The gin (processing plant) operates a fleet of trucks to transport the modules as required for operations during the ginning season. Farmers are paid for the seed cotton that crosses the scale at the gin. The gin operates a warehouse and stores bales of ginned cotton for periodic delivery to provide a year-round supply for their textile mill customers.

Figure 7. Module hauler picking up module stored at edge of field.

The advantage of the cotton model is that the specialized equipment used for hauling the modules is owned, scheduled, and managed by the gin. Thus, the producer does not own a piece of equipment they will use only a few times a year. The gin deploys the module haulers to their customers, thus the haulers accumulate more hours per years over what a producer would use, and the hauling cost ($/Mg) is minimized [40]. Typically, gins haul modules in the order they are "called in" by the producers. Module storage time in the field, and any subsequent losses (at the gin), are not dealt with in the producer-gin contract. The producer is paid the contract price for the mass of cotton fiber (and co-products) that the gin produces from a particular module.

7.2.3. Sugarcane (Texas model)

The producer grows the sugarcane. The sugar mill takes ownership of the crop in the field and harvests and transports it for continuous operation during the processing season. Sugarcane must be processed within 24 hours of harvest so storage is not an option within the sugarcane system.

The advantage of the Texas model is that the producer does not own any harvesting or hauling equipment. The disadvantage of the system is the producer does not have control over the time of harvest. Sugar content peaks (point of maximum compensation) in the middle of the season and a producer who has their cane harvested early, or late, sells less sugar. Fortunately, the issue is addressed in their contract.

8. Development of concept for multi-bale handling unit

8.1. Modulization of bales

Individual handling of bales (round or square) is not cost effective because of long loading and unloading times. Several concepts for a multi-bale handling unit are being developed. A verified multi-bale system is not available at this time to use as an example.

Permission was received [67] to present a concept that is far enough along in development that some field tests have been done. The concept was developed by a consortium led by FDC Enterprises and is shown in Figure 8. The self-loading trailer loads six stacks of six large rectangular bales, referred to "6-packs," for a total load of 36 bales. The bale size is 0.91

m × 1.22 m × 2.44 m. The length of the load is 6 × 2.44 = 14.6 m (48 ft), the height is 3 × 0.91 = 2.74 m (9 ft), and the width is 2 × 1.22 m = 2.44 m (8 ft).

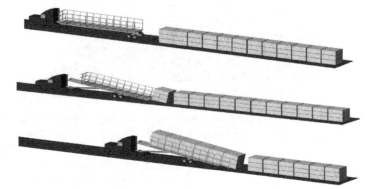

Figure 8. Multi-bale handling unit concept developed for 3x4x8 large rectangular bales by FDC Enterprises-led Consortium (Reprinted with permission [67]).

The trailer built to implement the concept [68] is shown in Figure 9. Estimated load time is 5 min, which is about the same load time for the cotton module (Figure 7). The 36-bale unit can be off-loaded by the truck directly onto the conveyor into a bioenergy plant, or it can be off-loaded into at-plant storage to be used later, just as is done with cotton modules at a cotton gin.

A similar concept known as the "Rack System" envisions that round bales will be loaded into a rack in-field or at an SSL. This rack is lifted off a trailer at the plant, emptied as the bales are needed, and then returned to be refilled. The racks are cycled multiple times each week within the closed logistics system. Cundiff et al. [59] developed the rack system concept, and it will now be used as an example to illustrate how the principles required for a logistics system are implemented.

All the various concepts and options cannot be discussed. However, we will apply a specific example to help the reader think through a process of developing a logistics system. The selection of the "Rack System" for this example implies no criticism of other ways of implementing the multi-bale handling unit concept currently being developed.

Figure 9. Self-loading trailer built by Kelderman Mfg. to implement multi-bale handling unit concept for 36-bale stack of large rectangular bales.

8.1.1. Receiving facility

When a truckload of feedstock arrives at the bioenergy plant, what happens? Does this material go directly into the plant as in a just-in-time processing model? Or, does it go into some type of at-plant storage? If it goes into at-plant storage, is there an efficient procedure for placing it into storage and retrieving it? All these questions relate directly to the cost of Receiving Facility operations, thus, when a bioenergy plant commissions a logistics system, it first specifies how it wants to receive material. Key questions for a receiving facility are:

1. Do I want each load to be the same size with the truck loaded the same way (i.e. are bales need to be in the same configuration)?
2. What are the hours of operation, and can I schedule approximately the same number of deliveries each hour of the workday?
3. What size at-plant storage must be maintained?

Design of a receiving facility is beyond the scope of this chapter; but, it is hoped that by posing these questions, the reader will begin to think about the constraints that a bioenergy may have when specifying a logistic system. Two criteria used in this example are:

1. Weigh and unload a truck in 10 min, and
2. Cost effective flow of material in-and-out of at-plant storage to support 24/7 operation.

Design of a logistics system is directly linked to the design of the receiving facility – neither is designed independent of the other.

8.1.2. Farmgate contract

Creation of a multi-bale handling unit will require specialized equipment. It is not a cost-effective to require each farmgate contract holder to own this equipment The rack system, for example, envisions a farmgate contract whereby the contract holder grows the crop, harvests in round bales, and places these bales in storage at a specified SSL. The contract holder owns and maintains the SSL and is paid a storage fee for each unit of feedstock that is stored. The biomass is purchased by the bioenergy plant in the SSL. All agricultural operations are now "sequestered" in the farmgate contract which gives those seeking a farmgate contract a well-defined process to prepare their bid.

A SSL must be graded to a minimum slope specification, must be near a main road, and must have a compacted gravel base. The gravel reduces bale degradation from water damage on the bottom of the bale and provides a suitable surface for equipment operations. Each SSL will receive material from either a single large production field or multiple, smaller fields as proposed by Cundiff et al. [59].

8.1.3. Hauling contract

The rack system envisions that the hauling contractor will invest in industrial equipment needed for year-round operation. Because the hauling contractor is hauling year-round, they can a) afford to invest in higher capacity industrial-grade equipment designed for up to 5,000 hour/year (or more) operation, and b) their labor force will develop expertise at the operations, and the Mg handled per unit of equipment investment will be a maximum. These two factors together create the potential to minimize hauling cost ($/Mg).

8.1.4. Storage

The logistics system has three storage features. Round bale packages act as self-storage and protect the biomass. Rounded top sheds water, so the round bale can be left in the field for a short time before in-field hauling. This provides the advantage of uncoupling the harvest and in-field hauling operations, and thus provides an opportunity for improving the cost efficiency of both operations. The farmgate contract holder has the opportunity to bale when the weather is right and haul later.

The second storage features are the SSLs. These locations provide the needed transition between in-field hauling and highway hauling. The SSLs will be located so that the Mg-km parameter for each SSL will be not more than 4 km. This means that, averaged across all Mgs stored at that SSL, each Mg will be hauled less than 4 km from the production field to the SSL. This constraint gives the producer an upper bound for calculating in-field hauling cost and all farmgate contractors are treated the same relative to in-field hauling cost.

The third storage feature is the inventory in at-plant storage, which provides the needed feedstock buffer at the plant. Those building a bioenergy plant would like to operate with just-in-time (JIT) delivery of feedstock as this gives them the lowest cost for receiving facility operation. If JIT is not possible, they want the smallest at-plant inventory for cost effective operation. There is obviously a trade-off in the logistics system design between the higher cost to purchase JIT delivery, and the cost of at-plant storage operations.

8.1.5. At-plant storage

In day-to-day management, it will be very difficult to achieve JIT delivery of any feedstock for 24/7 operation. All multi-bale handling unit concepts must include some at-plant storage. Even when known quantities of feedstock are stored in a network of SSLs, and a Feedstock Manager is controlling the deliveries, there will be delays.

To give a frame of reference, suppose 3 days of at-plant storage is the design goal. This inventory amount may suffice in the Southeast where ice and snow on the roads is not typically a significant problem for winter operations. But in the Midwest, additional days of at-plant storage will be required. A visualization of 3-day at-plant storage is shown in Figure 10. The number of racks shown is not part of the "cost analysis" example given later.

Bales will remain in the rack until processed; there is no individual bale handling at the plant. This is a very important aspect of any multi-bale handling system. This reduction in bale handling not only reduces cost, but also reduces damage to the bales. The integrity of the bales will be maintained, and bales will have additional protection from the rain since the rack has a top cover.

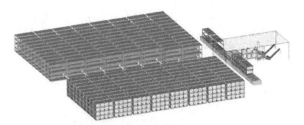

Figure 10. Illustration of at-plant storage for the rack system. At-plant storage shown with 144 racks with bales, 5 racks on conveyor (2 full, 1 being emptied, 2 empty), and 176 empty racks.

8.2. Functionality analysis for rack system concept

In this example, the analysis begins with round bales in ambient storage in SSLs and ends with a stream of size-reduced material entering the bioenergy plant for 24/7 operation.

8.2.1. List of baseline constraints

The constraints for this example are:

1. The rack design, and the design of the trailer to haul the rack, must conform to the standards for highway transport.
2. The rack used for this example holds sixteen 5×4 round bales. This round bale has a diameter of 1.52-m (5 ft) and a length of 1.22-m (4 ft). With two racks on a truck, a truckload is 32 this kind of round bales.
3. The bales will be pre-loaded into the racks while the trailer is parked at the SSL. When a truck arrives, a trailer with two empty racks is dropped, and the trailer with two full racks is towed away. The goal for the load time (trailer exchange) is 10 min, or less.
4. Hauling of the trailer and racks will be done 24 hours/day. Truck drivers will work five, 8-hour shifts per week and operations will be continuous for a 6-day work week. The rack loading crew at an SSL will be organized such that each worker will work a 40-h week. SSL operations will load bales into racks six 10-h workdays per week.
5. At the Receiving Facility, the racks are lifted from the trailer and placed on a conveyor to be conveyed into the plant for immediate use, or stacked two high in at-plant storage. The goal at the Receiving Facility is for the forklift to lift two full racks from the trailer and replace them with empty racks within of 10 min, or less.
6. The plant will operate with a maximum of 3 days of at-plant storage.
7. The example assumes that the plant processes one bale per minute. Assuming a bale weighs 900 lbs (0.408 Mg) in average (15% M.C.). The dry mass (DM) per bale is,

$$0.408 \ Mg \ / \ bale \ \times \ (1-0.15) \ = \ 0.3468 \ Mg \ DM \ / \ bale$$

Assuming one bale is processed per minute, the processing rate is,

$$0.33468 \ Mg \ DM \ / \ bale \ \times \ 1 \ bale \ / \ min \ \times \ 60 \ min \ / \ h \ = \ 20.8 \ Mg \ DM \ / \ h$$

For comparison, 20.8 Mg DM/h is 500 Mg DM/d. This size is in the 100 to 1000 Mg/d range recommended for Regional Biomass Processing Depots by Eranki et al. [54].

The processing time for racks is:

$$\frac{60 \text{ bale/h}}{16 \text{ bale/rack}} = 3.75 \text{ rack/h}$$

Thus, the number of truckloads consumed by the plant for a 24/7 operation is,

$$3.75 \, rack/h \times 24 \, h/d \times 7 \, d/wk = 630 \, rack/wk = 315 \, truck \, loads/wk$$

8.2.2. Operation Plan for 24-h Hauling

The plant will operate 24/7, but the Receiving Facility will be open 24 hours per day for 6 days. At 6:00 Sunday morning, there will be enough feedstock accumulated in at-plant storage for the maximum 3-day buffer (72 hours). Hauling operations begin again at 6:00 Monday morning. At-plant inventory is decreased to a 2-day buffer (48 hours) to supply the material for operation from 0600 Sunday morning to 6:00 Monday morning.

To discuss the 24-h-hauling concept, it is convenient to consider SSL operations as: 1) "day-haul" operation, and 2) "night-haul" operations. For the day-haul operation, the racks are transported as they are filled during the day. For the night-haul operation, the required number of empty racks, enough for one day's operation, are pre-positioned at the SSL. Cost of pre-positioning the racks is not considered in this example. The SSL crew fills these racks with bales during their 10-h workday and they are hauled during the night. Each truck arriving during the night unhooks a trailer with two empty racks and hooks up a trailer with two full racks. The next morning, the SSL crew will fill the empty racks delivered during the night and fill them during their workday.

8.2.3. Operational plan for receiving facility

A forklift (10-ton or 9.07-Mg capacity) will operate continuously at the receiving facility. This forklift will lift two full racks from the trailer and place them onto a conveyor into the plant for direct processing, or stack these racks in at-plant storage. Then the forklift will lift two empty racks onto the trailer and then the truck returns to the SSL. Empty racks will be stacked in the storage yard until they are lifted onto trailers.

The operational plan calls for two forklifts at the receiving facility, identified as a "work horse" and a "backup." The workhorse will operate continuously and the backup will operate during the day when trucks are waiting in the queue. Key point – the system must have a backup forklift because, if a forklift is not available to lift racks off of and onto trailers, all operations cease.

The handling of the racks emulates the handling of bins at a sugar mill in South Florida. In the bin system, a truck has three bins, two on the first trailer and one on a "pup" trailer. The bins are side-dumped if material is processed directly (Figure 11), or the bins are off-loaded and stacked two-high in the storage yard for nighttime operation (Figure 12). When the bins are dumped directly, it takes 3 min to dump the bins. For normal operation, one truck hauls 10 loads (30 bins) a day. At 37 tons/load (33.6 Mg/load), each truck hauls 370 ton/d (336 Mg/d). Sugar cane is 80% moisture content, so 370 ton = 74 dry ton/d/truck (67.2 Mg DM/d/truck).

Figure 11. Bins being side-dumped at sugar mill in South Florida, USA.

The conveyor for moving the racks into the plant is an adaptation of a piece of commercial technology. Conveyor design and function is similar to the conveyor used to move cotton modules into a cotton gin. Cotton modules and the 16-bale racks have approximately the same dimensions. The conveyor cost used in the analysis was obtained from a cotton equipment manufacturer. The receiving facility operates 6 d/wk, thus, on average, the daily delivery will be:

$$\frac{630 \text{ racks/wk}}{6 \text{ d/wk}} = 105 \text{ racks/d} = 53 \text{ trucks/d}$$

For this example, plant size (23 dry ton/h or 20.9 Mg DM/h) was chosen based on the expected capacity of the two forklifts at the plant. One forklift is expected to lift two full racks and lift and load two empty racks on a trailer at the rate of one truck every 27 minutes averaged over the 24-h a day. The design of the receiving facility and at-plant storage area has to facilitate this operation. A larger at-plant storage area will lower the forklift productivity (ton/h) because the average cycle time to move an individual rack is greater.

8.2.4. Size of at-plant storage yard

For example, maximum at-plant storage for a 3-day supply is, 3.75 racks/h times 72 h, or 270 racks. Racks will be stacked two high in "units" with two rows of 24 spaces each (Figure 12), thus there are 48 storage spaces in each unit. Each unit stores 96 racks. Three units are required for 270-racks storage.

$$3.75 \text{ racks} / h \quad x \quad 72 \text{ } h = 270 \text{ racks}$$

If 7-day at-plant storage is required, the total number of racks increases to 630 racks, or seven 96-rack units as shown in Figure 14. This implementation of a 7-day supply is not believed to be a cost effective choice, not only because of the capital investment in the racks, but also because of the forklift operating time required to cycle back and forth over a storage area this large. This is a key issue---the larger the at-plant storage, the lower the forklift productivity (ton/h), and thus the higher the forklift cost ($/ton).

The rack system competes best when the racks are filled and emptied as many times as possible in a given time period, not when they act as storage units. Other multi-bale handling units are more suitable for a larger at-plant storage like the one shown in Figure 14.

Figure 12. Bins being stacked in at-plant storage at sugar mill in South Florida, USA.

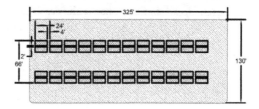

Figure 13. Sample layout of at-plant storage showing a unit (48 storage spaces). Racks are stacked two high for a total capacity of 96 racks (Total area required is approximately 4047 m²).

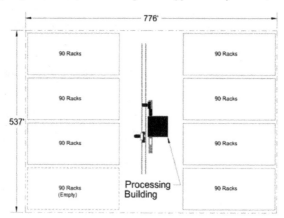

Figure 14. Sample layout of at-plant storage showing a 7-day supply of full racks and space for storing one unit of empty racks. (Total area required is approximately 38445 m²).

8.2.5. Bale loading into racks at the SSL

Ideally, the bale- loading- into- racks operation at the SSL should be able to load 16 bales in a rack in 20 minutes. This design criterion has not been attained with actual equipment.

For a workday with 10 productive hours (600 min), a 20-min-per-rack operation can theoretically load 30 racks, or 15 truckloads. Given the reality of SSL conditions, an actual operation could not sustain this productivity. In this analysis, it is reasonable to assume that a mature operation can average 70% of the theoretical productivity. The number of loads-day-operation used for this analysis will be reduced to 15×0.7 = 10.5 loads/d. The number of individual SSL operations required is 315 loads per week, divided by 6-day SSL operation per week, which is 52.5 or 53 loads per day average operation. This is five operations averaging 10.5 loads per day.

Thus, loading operations will occur at five different SSLs with a different set of equipment for each operation. Each SSL operation will load an average of 21 racks per day. Time to move the crew, equipment and reposition the empty trailer/rack from one SSL to the next is not dealt with in this analysis. Thus, the 21 racks/d productivity may be optimistic.

There are several options to load the rack. The option which found to be the most cost effective is the "side-load" option [58]. A telehandler with special attachment can pick up two bales per cycle (Figure 15) and load these bales into the side of the rack. Assuming that the average productivity to be achieved under production conditions is 34 min per rack, then the time needed to load two racks on a trailer is 68 min. Remember, this is the assumed average load time for year-round operation.

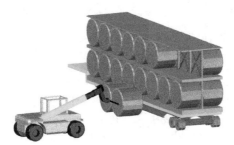

Figure 15. Concept for side loading bales into rack on trailer.

At the SSL, loading of the racks is the most challenging design for a cost-effective biomass logistics system. It is difficult to reduce the cost of these operations because the labor productivity (tons handled per worker-hour) might be low. By uncoupling the SSL loading and highway hauling, the truck does not have to wait for racks to be loaded in order to pick up the trailer. Also, the SSL crew does not have to wait for a truck to arrive with empty racks. However, the hooking/unhooking of trailers may be problematic for some drivers.

The day-haul operations are uncoupled by providing two extra trailers at the "day haul" SSL, and the night-haul operations are uncoupled by providing 9 extra trailers for the "night haul' SSL. Each truck tractor then has 11 extra trailers (22 racks) in the system. This may not be a best (least cost) approach, but it does provide a reasonable starting point.

8.2.6. Influence of SSL Size on Rack Loading Operations

If the yield of switchgrass in a commercial-scale operation averages 4 ton/ac (8.96 Mg/ha), this equates to approximately 9 bales/ac (22 bales/ha). It is suggested that the minimum size of a SSL contain the bales from 60 ac (24.5 ha) is 540 bales, and the maximum size SSL approximately stores bales from 1,200 ac (487 ha) 10,800. If 70% of the theoretical loading (30 racks/d) is achievable, the contactor will load 21 racks times 16 bales/rack is equal to 336 bales per day. The minimum size SSL will remove all of the bales in about 540/336, which is 1.6 days. The maximum size SSL will be loaded out in 10,800/336 = 32 days.

Cost of SSL operations at the smaller SSLs will be higher because of the mobilization charge to move equipment and crew to the next SSL for a fewer days of operation. It is probable that the load-haul contract offered by the biorefinery for each individual SSL will consider the haul distance and SSL size and thus the per-ton price will be different for each SSL.

8.2.7. Total Trucks Required – 24-h Hauling

To achieve 24-h hauling, the truck drivers will work 8-h shifts and the trucks will run continuously from 0600 Monday to 0600 Sunday, a total of 144 h/wk. The total racks processed each week are 630, equal to 315 truckloads. If a uniform delivery is assumed, the average truck unload time is,

$$315 \; trucks \; / \; 144 \; h \; = \; 2.2 \; trucks/h \; = \; 1 \; truck/27 \; min.$$

This productivity is well within the Rack System design goal of a 10 min unload time. As previously stated, the 24-h hauling concept envisions that the SSL crew will leave a supply of loaded racks on trailers at the SSL when they finish their 10-h workday. These racks will be hauled during the night. The next morning, the loading crew will load empty racks delivered during the night and load them during their workday.

The key variable in hauling is the truck cycle time. To calculate cycle time, for example, an average haul distance is needed. An actual database was developed for a proposed bioenergy plant location at Gretna, Virginia, USA to calculate the average haul distance.

An analysis was done for a 30-mi (48 km) radius around Gretna to identify potential production fields based on current land use determined from current aerial photography and GIS methods. It is conservatively assumed that 5% of the total land was assigned into switchgrass production. SSLs were established at 199 locations (Figure 5), and the existing road network was used to determine the travel distance from each SSL to the proposed plant location at Gretna. Some loads were hauled 2 miles (3.2 km) and some were hauled over 40 miles (64 km). A weighted ton-mi parameter was computed and found to be 25.4 miles (40.6 km). This means that, across all 199 SSLs, each ton travels an average of 25.4 miles (40.6 km) to get to the plant.

Truck cycle time is calculated using the 25.4-mile average haul distance, a 45 mph average speed, 10-min to hook/unhook trailers a SSL, and 10-min to lift full and empty racks from the trailers. Theoretical cycle time is 1.46 h. In 24 hours of operation, one truck can haul:

$$24 \, h / (1.46 \, h / load) \; = \; 16.4 \; loads \; per \; truck \; per \; day$$

Assuming that a truck fleet can average 70% of the theoretical capacity, then (0.7×16.4 =11.5 loads per truck per day can be achieved. Remember, since the trucks run continuously, a decimal number of loads can be used as the average achieved per-day of productivity.

It is not practical to use the each-haul contractor-runs-their-own-trucks assumption for 24-h hauling. The way to maximize truck fleet productivity is to have the Feedstock Manager have the control to send any truck to any SSL where a trailer with full racks is available. This greatly facilitates the hauling at both day-haul and night-haul SSLs.

Total trucks being controlled by the Feedstock Manager is:

$$53 \; loads / day \; required \; at \; the \; plant \, / \, 11.5 \; loads \; per \; truck \; = \; 4.6 \; trucks \; (5 \; trucks)$$

Since five SSL crews, one per SSL, are loading trailers/racks. More realistic is that eight to ten trucks will be available and some would also have responsible for bring other supplies to the biorefinery or hauling waste and other value-added products from the biorefinery.

8.2.8. Total Racks and Assets Required for 24-h Hauling

Since the only time deliveries are not being made is the 24-h period from 0600 Sunday to 0600 Monday, the amount in at-plant storage can be reduced. Using 1.5 days as the minimum at-plant storage, so the total capacity hours required in at-plant storage at 0600 Sunday, when deliveries are ended for the week, is

$$24 \, h \; (actual) \; + \; 1.5 \, d \, \times \, 24 \, h / d \; (at - plant \; storage) \; = \; 60 \; hours$$

$$3.75 \; racks / h \; \times \; 60 \, h \; = \; 225 \; racks$$

Total trailers are calculated as follows. Each truck has one trailer connected, two parked at a "day-haul" SSL and nine parked at a "night-haul" SSL for a total of 12 trailers. The total racks on trailers are calculated as:

$$5 \; trucks \; \times \; 12 \; trailers \; per \; truck \; \times \; 2 \; racks / trailer \; = \; 120 \; racks$$

Total racks required are:

At-plant	+	On 60 trailers	+	Reserve	=	Total
225	+	120	+	5	=	350

The actual number of racks required is calculated by subtracting the racks on parked trailers from the rack total (empty + loaded) at the plant. Potentially, 60 loaded trailers can be parked at the receiving facility when hauling ends for the week at 0600 Sunday. In order for this procedure to work, the racks on most of these 60 trailers have to be returned to SSLs during the period 0600 Sunday to 0600 Monday so they will be in position for operations to begin at each SSL at 0600 Monday. This requires some empty back hauls. Cost for these empty back hauls is a level of detail that must wait for a more sophisticated analysis.

When racks on trailers are counted as part of the at-plant storage, the minimum number of racks is:

At-plant	+	On 60 trailers	+	Reserve	=	Total
(225-120)	+	120	+	5	=	230

Average number of cycles per rack is 29,610 racks processed per year divided by 230, that is 129 cycles per year; or about 2.7 cycles per week for 47 weeks of annual operation.

8.3. Cost analysis for 24-h hauling using rack system

The costs given in this section are presented without supporting detail. They were calculated using the procedures given in [69]. They are "best estimates" given current cost parameters. All costs are given on a $/Mg DM basis for operation of a bioenergy plant consuming 23 dry ton/h (20.9 Mg DM/h). The challenge is to find a way that machine productivity (Mg/h) can be increased.

8.3.1. Total truck cost

The assumed truck cost (tractor and trailer for hauling the two racks) is $630/d for a 24-h workday, which includes ownership plus operating cost, plus labor, but excluding fuel.

Truck cost, excluding fuel, is:

$$\frac{\$630/d}{11.5 \text{ loads/d} \times 12.2 \text{ dry ton/load}} = \$4.49/\text{dry ton} = \$4.95/\text{Mg DM}$$

Truck fuel cost for the 25.4 miles (40.6 km) average haul distance is:

$$(25.4 \text{ mi} \times 2)/4 \text{ mi/gal} = 12.7 \text{ gal} \times \$3.50/\text{gal} = \$44.45/\text{load}$$

$$\$44.45/\text{load} / 12.2 \text{ dry ton/load} = \$3.64/\text{dry ton} (\$4.01/\text{Mg DM})$$

Total truck cost is ownership and operating plus fuel, that is 4.95 plus 4.01 = $8.96/Mg DM.

8.3.2. Load, unload operations

1. Handling racks at plant: $1.93 (forklift) + $1.02 (backup forklift) = $2.95/dry ton
2. SSL operation: $3.66 (telehandler) + $0.98 (extra trailers) = $4.64/dry ton
3. Rack cost: cost for 230 racks = $1.80/dry ton

4. Storage yard at processing plant: $0.13/dry ton
5. Conveyor entering plant: $0.28/dry ton

(Note: 1 dry ton = 0.907 Mg DM)

8.3.3. Size Reduction

1. Unroller-chopper cost: $5.76/dry ton.
2. Rack cost: all costs associated with the ownership and maintenance of the racks.
3. Loading cost: all costs associated with the loading of bales into racks. These costs are referred to as "SSL operation costs".
4. Truck cost: all costs associated with the ownership and operation of the trucks.
5. Receiving Facility cost: all costs associated with the unloading of racks from trucks, placement of racks onto conveyor (or placement in at-plant storage), conveyor operation, operation of at-plant storage, and removal of racks from at-plant storage and placement on trucks for return to SSL.
6. Size reduction: all costs associated with the unloading of bales from the rack, operation of conveyor for single file bales delivered to size reduction machine, and operation of machine for initial size reduction.

Truck cost is 34% of the total cost, SSL operations are 20%, Receiving Facility operations are 14%, size reduction is 24%, and the racks are 8%. It is clear why the Rack System Concept was organized to maximize truck productivity; truck cost is the largest cost component. Truck cost plus SSL operations are $12.77/dry ton, or 54% of total cost. The Receiving Facility cost is $3.37/dry ton, only 14% of total cost. As with all other multi-bale handling system concepts, the Rack System provides an opportunity for minimizing cost between the plant gate and the size reduction unit operation.

The total cost shown in Table 2 does not include the farmgate contract cost (production, harvesting, in-field transport, storage in SSL, and profit to producer). The farmgate contract cost can be estimated from local data for production, harvest, and ambient storage of round bales of hay. In the Southeast the key issue relative to the hay cost comparison is the difference in yield; switchgrass will yield about 9 Mg/ha as compared to traditional hay species that yield about 4.5 Mg/ha.

Operation	Cost ($/dry ton)*
Racks	1.80
Loading at SSL	
Telehandler	3.66
Extra Drop-deck Trailers	0.98
Truck cost	8.13
Unloading at plant	
Workhorse forklift	1.93
Backup forklift	1.02
At-plant storage (Gravel lot with lighting)	0.13

Operation	Cost ($/dry ton)*
Conveyor into plant	0.28
Unroller-chopper (Initial size reduction)	5.76
Total	$23.69

* $1/dry ton = $1.103/Mg DM

Table 2. Total cost for hauling, receiving facility operations, and size reduction for rack system example – side-load option, 24-h hauling.

9. Conclusions

The key decision points for the design of a logistics system for a bioenergy plant operating 24/7 year-round are summarized as follows.

1. A complete logistics system is defined as one that begins with the biomass standing in the field and ends with a stream of size-reduced material entering a bioenergy plant for 24/7 operation. Optimizing one unit operation in isolation may increase the cost of an "upstream" or "downstream" operation such that total delivered cost is increased.
2. Most feedstock is harvested only part of the year, thus storage is a part of the logistics system. A cost effective logistics system provides for efficient flow of material in and out of storage.
3. Just-in-time (JIT) delivery of feedstock provides for a minimum at-plant storage cost. Since JIT delivery is not practical for typical biomass logistics systems, there is always a cost trade-off between the size of at-plant storage and the other design constraints needed to insure a continuous feedstock supply. Knowledge of quantities of biomass in Satellite Storage Locations (SSLs) provides the Feedstock Manager at a bioenergy plant an opportunity to minimize the at-plant storage cost.
4. Farmgate contracts that require a summer-early fall harvest must compensate for the removal of nutrients, and contracts that require a winter harvest must compensate for loss of yield incurred by the delayed harvest.
5. Assigning different unit operations to different entities in the business plan can lower average delivered cost. For example, it is more efficient to pool all farmgate activities into a farmgate contract and have a hauling contractor handle all load-haul activities. This division is defined as a division between "agricultural" and "industrial" operations. The key benefit achieved is in the capitalization of the equipment. Load-haul contractors can afford to invest in industrial-grade, high-capacity equipment designed for year-round operation as compared to farmgate contractors who will use their equipment 400 hours (or less) per year.
6. Uncoupling of the unit operations in the logistics chain can provide an advantage.
    a. In the agricultural operations, baling uncouples the harvesting and in-field hauling operations. When the harvesting operation is not constrained by in-field hauling--- both unit operations can proceed at maximum productivity.

b. In the industrial operations, it is important to uncouple truck loading from hauling. Maximum loads-per-day-per-truck are achieved when the loading crew never has to wait for a truck to arrive and the truck never has to wait to be loaded.

7. Truck cost is the largest component of total cost in most logistics systems, thus it is essential to maximize truck productivity (Mg hauled per unit time) by increasing both Mg-per-load and loads-per-day. A 10-min load time and a 10-min unload time is a desired goal for design of most logistic systems.

8. Multi-bale handling units are in need to solve the rapid loading/unloading challenge.

9. Twenty-four-hour hauling can minimize truck cost ($/Mg). The challenge is to design a logistics system with a practical procedure for loading trucks at night at a remote location.

10. The design of the Receiving Facility, because of the need to unload trucks quickly, is critical in the design of a complete logistics system. Typically, this design specifies that each load have the same configuration, and requires a delivery schedule where approximately the same number of loads is received each workday.

11. The most cost-effective logistics system will be structured such that information technologies (GPS, bar codes, entry of data over cell phone network) and optimization routines developed for other logistics systems can be used to optimize asset utilization in real time.

Author details

Jude Liu*
Department of Agricultural and Biological Engineering, The Pennsylvania State University, University Park, Pennsylvania, USA

Robert Grisso and John Cundiff
Department of Biological Systems Engineering, Virginia Tech, Blacksburg, Virginia, USA

10. References

[1] Brownell D K, Liu J, Hilton J W, Richard T L, Cauffman G R, and MacAfee B R (2009) Evaluation of two forage harvesting systems for herbaceous biomass harvesting. ASABE Paper No. 097390, St. Joseph, MI: ASABE.

[2] Reynolds S G, and Frame J (2005) Grasslands: Developments, opportunities, perspectives. Enfield, New Hampshire: Science Publ. Inc.

[3] Cundiff J S, and Grisso R D (2008) Containerized handling to minimize hauling cost of herbaceous biomass. Biomass & Bioenergy 32(4): 308-313

[4] Hierden E V (1999) Apparatus for processing large square hay bales into smaller recompressed Bales. U.S. Patent No. 6339986.

* Corresponding Author

[5] Steffen Systems (2009) High Density Bale Compression Systems. Available at: http://www.steffensystems.com/Products/Bale_Press/index.htm. Accessed 2012 April 19.

[6] Sokhansanj S, Turholow A F, Stephen J, Stumborg M, Fenton J and Mani S (2008) Analysis of five simulated straw harvest scenarios. Can. Biosys. Eng. 50:2.27-2.35

[7] Sokhansanj S, Turhollow A F and Wilkerson E G (2008) Development of the Integrated Biomass Supply Analysis and Logistics Model (IBSAL). Technical Memorandum ORNL/TM-2006/57. Oak Ridge, TN: Oak Ridge National Laboratory.

[8] Kumar, P.K., and K.E. Ileleji. 2009. Tech-no-economic analysis of the transportation, storage, and handling requirements for supplying lignocellulosic biomass feedstocks for ethanol production. ASABE Paper No. 097427, St. Joseph, MI: ASABE.

[9] Brownell D K (2010) Analysis of biomass harvest, handling, and computer modeling. M.S. Thesis. The Pennsylvania State University, University Park, PA.

[10] Srivastava A K, Carroll E G, Rohrbach R P, and Buchmaster D R (2006) Engineering principles of Agricultural Machines. American Society of Agricultural and Biological Engineers, St. Joseph, MI: ASABE.

[11] Cundiff J S, Grisso R D, and McCullough D (2011) Comparison of bale operations for smaller production fields in the Southeast. ASABE Paper No. 1110922. St. Joseph, MI: ASABE.

[12] ASAE Standards D497.5 (2006) Agricultural machinery management data. American Society of Agricultural Engineers, St. Joseph, MI: ASAE.

[13] Liu J and Kemmerer B (2011) Field performance analysis of a tractor and a large square baler. SAE Technical Paper 2011-01-2302, Presented at Commercial Vehicle Engineering Congress, Chicago, IL doi:10.4271/2011-01-2302.

[14] Kemmerer B, and Liu J (2012) Large square baling and bale handling efficiency—A case study. Ag. Sci. 3(2):178-183.

[15] ASAE Standards EP496.3 (2006) Agricultural machinery management. American Society of Agricultural Engineers, St. Joseph, MI: ASABE.

[16] Cundiff J S, Fike J H, Parrish D J, and Alwang J (2009) Logistic constraints in developing dedicated large-scale bioenergy systems in the Southeastern United States. Journal of Environmental Engineering-ASCE 135(11):1086-1096.

[17] Hadders G, and Olsson R (1997) Harvest of grass for combustion in late summer and in spring. Biomass Bioenergy 12(3):171–175.

[18] Larson J A, Mooney D F, English B C, and Tyler D D (2010) Cost analysis of alternative harvest and storage methods for switchgrass in the Southeastern U.S. 2010 Annual Meeting of Southern Agricultural Economics Association, Feb. 6-9, 2010, Orlando, FL.

[19] Epplin F M, Mapemba L, and Tembo G (2005) Economic modeling of a lignocellulosic biomass biorefining industry. In: Outlaw JL, Collins KJ, Duffield JA, editors. Agriculture as a producer and consumer of energy. London: CABI Publishing; 2005. p. 205–217.

[20] Hwang S, Epplin F M, Lee B, and Huhnke R (2009) A probabilistic estimate of the frequency of mowing and baling days available in Oklahoma USA for the harvest of switchgrass for use in biorefineries. Biomass Bioenergy 33(8):1037-1045.

[21] Sanderson M A, Egg R P, and Wiselogel A E (1997) Biomass losses during harvest and storage of switchgrass. Biomass Bioenergy 12(2):107–114.

[22] Kumar A, and Sokhansanj S (2007) Switchgrass (Panicum vigratum, L.) delivery to a biorefinery using integrated biomass supply analysis and logistics (IBSAL) model. Bioresource Technology 98, 1033-1044.

[23] Rees D V H (1982) A discussion of the sources of dry matter loss during the process of haymaking. J. of Agric. Engng Res. 27:469–479.

[24] Coble C G, and Egg R (1987) Dry matter losses during hay production and storage of sweet sorghum used for methane production. Biomass 14:209–217.

[25] Bledsoe B L, and Bales B M (1992) Bale wrap evaluation. ASAE Paper No. 921573. St. Joseph, MI: ASAE.

[26] LaFlamme L F (1989) Effects of storage conditions for large round bales on quality of grass-legume hay. Can. J. of Animal Sci. 69:955–961.

[27] Turhollow A (1994) The economics of energy crop production. Biomass Bioenergy 6(3):229-241.

[28] Shinners K J, Boettcher G C, Muck R E, Wiemer P J, and Casler M D (2010) Harvest and storage of two perennial grasses as biomass feedstocks. Trans. of ASABE 53(2):359-370

[29] Chariton Valley RC&D (2002) Chariton Valley Biomass Project - Draft Fuel Supply Plan, United States Department of Energy, Contract Number: DE-FC36-96GO10148

[30] Styles D, Thorne F, and Jones M B (2008) Energy crops in Ireland: an economic comparison of willow and Miscanthus production with conventional farming systems. Biomass Bioenergy 32(5):407-421.

[31] Khanna M, Dhungana B, and Clifton-Brown J (2008) Costs of producing Miscanthus and switchgrass for bioenergy in Illinois. Biomass Bioenergy 32(6):482-493.

[32] Monti A, Fazio S, Lychnaras V, Soldatos P, and Venturi G (2007). A full economic analysis of switchgrass under different scenarios in Italy estimated by BEE model. Biomass Bioenergy 31(4):177-185.

[33] Duffy M (2007) Estimated costs for production, storage and transportation of switchgrass. Ames, IA: Iowa State University, Extension Report No. PM 2042.

[34] De Mol R M, Jogems M A H, Van Beek P, and Gigler J K (1997) Simulation and optimization of the logistics of biomass fuel collection. Neth J Agric Sci 45:219-228.

[35] Cundiff J, Dias N, and Sherali H D (1997) A linear programming approach for designing a herbaceous biomass delivery system. Bioresource Technology 59:47-55.

[36] Zuo M, Kuo W, and McRoberts K L (1991) Application of mathematical programming to a large-scale agricultural production and distribution system. J Oper Res Soc 42(8):639-648.

[37] Mapemba L D, Epplin F M, Huhnke R L, and Taliaferro C M (2008) Herbaceous plant biomass harvest and delivery cost with harvest segmented by month and number of harvest machines endogenously determined. Biomass Bioenergy 32(11):1016-1027.

[38] Kumar A, Sokhansanj S, and Flynn P C (2006) Development of a multicriteria assessment model for ranking biomass feedstock collection and transportation systems. Appl Biochem Biotechnol 129-132:71-87.

[39] Mukunda A, Ileleji K E, and Wan H (2006) Simulation of corn stover logistics from on-farm storage to an ethanol plant. ASABE Paper No. 066177, St. Joseph, MI: ASABE.

[40] Ravula P P, Grisso R D, and Cundiff J S (2008) Cotton logistics as a model for a biomass transportation system. Biomass & Bioenergy 32(4):314-325.

[41] Sokhansanj S, Kumar A, and Turhollow A F (2006) Development and implementation of integrated biomass supply analysis and logistics model (IBSAL). Biomass Bioenergy 30:838-847.

[42] Shastri Y, Hansen A, Rodriguez L, and Ting K C (2011) Optimization of Miscanthus harvesting and handling as an energy crop: BioFeed model application. Biol Eng 3(1):37-69.

[43] Shastri Y, Hansen A, Rodriguez L, and Ting K C (2011) Development and application of BioFeed model for optimization of herbaceous biomass feedstock production. Biomass Bioenergy 35:2961-2974.

[44] Gallis C T (1996) Activity oriented stochastic computer simulation of forest biomass logistics in Greece. Biomass Bioenergy 10:377-382.

[45] Nilsson D (2000) Dynamic simulation of straw harvesting systems: influence of climatic, geographical and biological factors on performance and costs. J. Agric. Engng Res 76:27-36.

[46] Freppaz D, Minciardi R, Robba M, Rovatti M, Sacile R, and Taramasso A (2004) Optimizing forest biomass exploitation for energy supply at a regional level. Biomass Bioenergy 26:15-25.

[47] Judd J D, Sarin S C, Cundiff J S, and Grisso R D (2010) Cost analysis of a biomass logistics system. ASABE Paper No. 1110466. St. Joseph, MI: ASABE.

[48] Tatsiopoulos I, and Tolis A (2003) Economic aspects of the cotton-stalk biomass logistics and comparison of supply chain methods. Biomass Bioenergy 24:199-214.

[49] Tembo G, Epplin F M, and Huhnke R L (2003) Integrative investment appraisal of a lignocellulosic biomass-to-ethanol industry. J. Agric. Res. Econ. 28:611-633.

[50] Morey V R, Kaliyan N, Tiffany D G, and Schmidt D R (2010) A corn stover supply logistics system. Applied Eng. in Agric. 26(3):455-461.

[51] Nagel J (2000) Determination of an economic energy supply structure based on biomass using a mixed-integer linear optimization model. Ecol. Engng 16:91-102.

[52] Gan J, and Smith C T (2011) Optimal plant size and feedstock supply radius: A modeling approach to minimize bioenergy production costs. Biomass Bioenergy 35(8):3350-3359.

[53] Dunnett A, Agjiman C, and Shah N (2007) Biomass to heat supply chains applications of process optimization. Transactions Institution of Chemical Engineers: Process Safety and Environmental Protection 85(B5): 419–429.

[54] Bruglieri M, and Liberti L (2008) Optimal running and planning of a biomass-based energy production process. Energy Policy 36(7):2430-2438.

[55] Leduc S, Schwab D, Dotzauer E, Schmid E, and Obersteiner M (2008) Optimal location of wood gasification plants for methanol production with heat recovery. Int'l J. of Energy Res. 32(12):1080-1091.

[56] Berruto R, and Busato P (2008) System approach to biomass harvest operations: simulation modeling and linear programming for logistic design. ASABE Paper No. 084565, St. Joseph, MI: ASABE.

[57] Resop J P, Cundiff J S, and Heatwole C D (2010) Spatial analysis to site satellite storage locations for herbaceous biomass in the Piedmont of the Southeast. Applied Eng. Agric. 27(1): 25-32.

[58] Judd J D, Sarin S C, and Cundiff J S (2011) An optimal storage and transportation system for a cellulosic ethanol bio-energy plant. ASABE Paper No. 109413., St. Joseph, MI: ASABE.

[59] Cundiff J S, Grisso R D, and Ravula P P (2004) Management system for biomass delivery at a conversion plant. ASAE/CSAE Paper No. 046169, St. Joseph, MI: ASABE

[60] Eranki, P.L., Bryan D. Bals, and Bruce E. Dale. 2011. Advanced regional biomass processing depots: a key to the logistical challenges of the cellulosic biofuel industry. Biofuels, Bioprod. Bioref. DOI:10.1002/bbb. Society of Chemical Industry and John Wiley & Sons, Ltd.

[61] Yoshioka T, Aruga K, Nitami T, Sakai H, and Kobayashi H (2006) A case study on the costs and the fuel consumption of harvesting, transporting, and chipping chains for logging residues in Japan. Biomass Bioenergy 30:342-348

[62] Ranta T (2005) Logging residues from regeneration fellings for biofuel production-a GIS-based availability analysis in Finland. Biomass Bioenergy 28:171-182.

[63] Ranta T, and Rinne S (2006) The profitability of transporting uncomminuted raw materials in Finland. Biomass Bioenergy 30:231-237.

[64] Ayoub N, Martins R, Wang K, Seki H, and Naka Y (2007) Two levels decision system for efficient planning and implementation of bioenergy production. Energy Conv. Man. 48:709-723.

[65] Cundiff J S, Shapouri H, and Grisso R D (2007) Economic analysis of two receiving facility designs for a bioenergy plant. ASABE Paper No. 076051, St. Joseph, MI: ASABE

[66] Brownell D, and Liu J (2012) Managing biomass feedstocks: selection of satellite storage locations for different harvesting systems. Agric. Engng Int'l CIGR Journal, 14(1): Manuscript No. 1886. 13 pp.

[67] Cromer K (2011) Personal communication. FDC Enterprises-led Consortium. Antares Group Inc., 57 S. Main St., Suite 506, Harrisonburg, VA 22801. Cooper, Sam, courtesy of Sugar Journal, P.O. Box 19084, New Orleans, LA 70179).

[68] Kelderman Mfg. to implement multi-bale handling unit.

Aquatic Biomass

Biofloc Technology (BFT): A Review for Aquaculture Application and Animal Food Industry

Maurício Emerenciano, Gabriela Gaxiola and Gerard Cuzon

Additional information is available at the end of the chapter

1. Introduction

The aquaculture industry is growing fast at a rate of ~9% per year since the 1970s [1]. However, this industry has come under scrutiny for contribution to environmental degradation and pollution. As a result, requirement for more ecologically sound management and culture practices remains fully necessary. Moreover, the expansion of aquaculture is also restricted due to land costs and by its strong dependence on fishmeal and fish oil [2,3]. Such ingredients are one of the prime constituents of feed for commercial aquaculture [4]. Feed costs represent at least 50% of the total aquaculture production costs, which is predominantly due to the cost of protein component in commercial diets [5].

Interest in closed aquaculture systems is increasing, mostly due to biosecurity, environmental and marketing advantages over conventional extensive and semi-intensive systems [6]. When water is reused, some risks such as pathogen introduction, escapement of exotic species and discharging of waste water (pollution) are reduced and even eliminated. Furthermore, because of high productivity and reduced water use, marine species can be raised at inland locations [6]. A classic example is the currently expansion of marine shrimp farms at inland location in USA, which allows local farmers market fresh never frozen shrimp in metropolitan locations with good profitability.

The environmental friendly aquaculture system called "Biofloc Technology (BFT)" is considered as an efficient alternative system since nutrients could be continuously recycled and reused. The sustainable approach of such system is based on growth of microorganism in the culture medium, benefited by the minimum or zero water exchange. These microorganisms (biofloc) has two major roles: (i) maintenance of water quality, by the uptake of nitrogen compounds generating "*in situ*" microbial protein; and (ii) nutrition, increasing culture feasibility by reducing feed conversion ratio and a decrease of feed costs.

As a closed system, BFT has primordial advantage of minimizing the release of water into rivers, lakes and estuaries containing escaped animals, nutrients, organic matter and pathogens. Also, surrounding areas are benefitted by the "vertically growth" in terms of productivity, preventing coastal or inland area destruction, induced eutrophication and natural resources losses. Drained water from ponds and tanks often contains relatively high concentrations of nitrogen and phosphorous, limiting nutrients that induce algae growth, which may cause severe eutrophication and further anaerobic conditions in natural water bodies. In BFT, minimum water discharge and reuse of water prevent environment degradation and convert such system in a real "environmentally friendly system" with a "green" approach. Minimum water exchange maintain the heat and fluctuation of temperature is prevented [7], allowing growth of tropical species in cold areas.

Currently, BFT has received alternate appellation such as ZEAH or Zero Exchange Autotrophic Heterotrophic System [8-10], active-sludge or suspended bacterial-based system [11], single-cell protein production system [12], suspended-growth systems [13] or microbial floc systems [14,15]. However, researches are trying to keep the term "BFT or Biofloc Technology" in order to establish a key reference, mainly after the book release "*Biofloc Technology – A Practical Guide Book*" in 2009 [16]. Moreover, BFT has been focus of intensive research in nutrition field as a protein source in compounded feeds. Such source is produced in a form of "biofloc meal", mainly in bioreactors [17]. In addition, the fast spread and the large number of BFT farms worldwide induced significant research effort of processes involved in BFT production systems [14].

The objective of this chapter is to review the application of Biofloc Technology (BFT) in aquaculture; and describes the utilization of biofloc biomass (also described in this chapter as "biofloc meal") as an ingredient for compounded feeds. An addition goal is to help students, researchers and industry to clarify the basic aspects of such technology, aiming to encourage further research.

2. History of BFT

According to [18], BFT was first developed in early 1970s at Ifremer-COP (French Research Institute for Exploitation of the Sea, Oceanic Center of Pacific) with different penaeid species including *Penaeus monodon*, *Fenneropenaeus merguiensis*, *Litopenaeus vannamei* and *L. stylirostris* [19,20]. Such culture system was compared with an "external rumen", but now applied for shrimp [21]. At the same period, Ralston Purina developed a system based on nitrifying bacteria while keeping shrimp in total darkness. In connection with Aquacop, such system was applied to *L. stylirostris* and *L. vannamei* both in Crystal River (USA) and Tahiti, leding considerations on benefits of biofloc for shrimp culture [22]. In 1980, a French scientific program 'Ecotron' was initiated by Ifremer to better understand such system. Several studies enabled a comprehensive approach of BFT and explained interrelationships between different compartments such as water and bacteria, as well as shrimp nutritional physiology. Also in 1980s and beginning of 1990s, Israel and USA (Waddell Mariculture Center) started R&D in BFT with tilapia and white shrimp *L. vannamei*, respectively, in which water limitation, environmental concerns and land costs were the main causative agents that promoted such research (Fig. 1).

Figure 1. Biofloc technology at Ifremer, Tahiti (A), Sopomer farm, Tahiti (B), Waddell Mariculture Center (C) and Israel (D) (Photos A and B: Gerard Cuzon; C: courtesy of Wilson Wasielesky; and D: courtesy of Yoram Avnimelech)

Regarding to commercial application of BFT, in 1988 Sopomer farm in Tahiti (French Polynesia) using 1000m² concrete tanks and limited water exchange achieved a world record in production (20–25 ton/ha/year with two crops) [22, 23]. On the other hand, Belize Aquaculture farm or "BAL" (located at Belize, Central America), probably the most famous case of BFT commercial application in the world, produced around 11-26 ton/ha/cycle using 1.6 ha lined grow-out ponds. Much of know-how of running worldwide commercial scale BFT shrimp ponds is derived from BAL experience. In small-scale BFT greenhouse-based farms, Marvesta farm (located at Maryland, USA), probably is the well-known successful indoor BFT shrimp farm in USA, can produce around 45 ton of fresh never frozen shrimp per year using ~570 m³ indoor race-ways [24]. Nowadays, BFT have being successfully expanded in large-scale shrimp farming in Asia, Latin and Central America, as well as in small-scale greenhouses in USA, South Korea, Brazil, Italy, China and others (Fig 2). In addition, many research centers and universities are intensifying R&D in BFT, mostly applied to key fields such as grow-out management, nutrition, BFT applied to reproduction, microbial ecology, biotechnology and economics.

Figure 2. Biofloc technology commercial-scale at BAL (A) and Malaysia (B), and pilot-scale in Mexico (C and D) (Photos A, B and D: Maurício Emerenciano; and C: courtesy of Manuel Valenzuela)

3. The role of microorganisms

The particulate organic matter and other organisms in the microbial food web have been proposed as potential food sources for aquatic animals [25]. In BFT, microorganisms present a key role in nutrition of cultured animals. The macroaggregates (biofloc) is a rich protein-lipid natural source available *"in situ"* 24 hours per day [14]. In the water column occurs a complex interaction between organic matter, physical substrate and large range of microorganisms such as phytoplankton, free and attached bacteria, aggregates of particulate organic matter and grazers, such as rotifers, ciliates and flagellates protozoa and copepods [26] (Fig 3). This natural productivity play an important role recycling nutrients and maintaining the water quality [27,28].

The consumption of biofloc by shrimp or fish has demonstrated innumerous benefits such as improvement of growth rate [10], decrease of FCR and associated costs in feed [9]. Growth enhancement has been attributed to both bacterial and algae nutritional components, which up to 30% of conventional feeding ration can be lowered due to biofloc consumption in shrimp [29]. In reference [9] was reported that more than 29% of daily food consumed for *L. vannamei* could be biofloc. In tilapia, in [30] was estimated that feed utilization is higher in BFT at a rate of 20% less than conventional water-exchange systems.

Also, consumption of macroaggregates can increase nitrogen retention from added feed by 7-13% [31, 32]. In this context, BFT has driven opportunities to use alternative diets. Low protein feeds and feeds with alternative protein sources different than marine-based products (i.e. fishmeal, squid meal, etc) have been successfully applied in BFT [28, 33-35], leading "green" market opportunities.

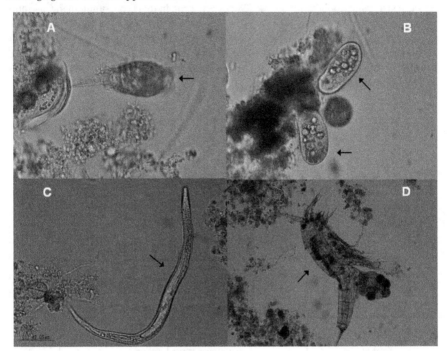

Figure 3. Grazers often observed in BFT such as flagellates protozoa (A), ciliates protozoa (B), nematodes (C) and copepods (D) (10x magnification) (Source: Maurício Emerenciano)

Regarding to maintenance of water quality, control of bacterial community over autotrophic microorganisms is achieved using a high carbon to nitrogen ratio (C:N) [30], which nitrogenous by-products can be easily taken up by heterotrophic bacteria [36]. High carbon to nitrogen ratio is required to guarantee optimum heterotrophic bacteria growth [14, 37], using this energy for maintenance (respiration, feeding, movement, digestion, etc), but also for growth and to produce new cells. High carbon concentration in water could supersede the carbon assimilatory capacity of algae, contributing to bacteria growth. Aerobic microorganisms are efficient in converting feed to new cell material (40-60% of conversion efficiency), rather than higher organisms that spend about 10-15% to rise in weight [16]. Bacteria and other microorganisms act as very efficient "biochemical systems" to degrade and metabolize organic residues [36]. In other words, they recycle very efficiently nutrients in a form of organic and inorganic matter (un-consumed and non-digested feed, metabolic residues and carbon sources applied as fertilizers) into new microbial cells.

The carbon sources applied in BFT are often by-products derived from human and/or animal food industry, preferentially local available. Cheap sources of carbohydrates such as molasses, glycerol and plant meals (i.e. wheat, corn, rice, tapioca, etc) will be applied before fry/post-larvae stocking and during grow-out phase, aiming to maintain a high C:N ratio (~15-20:1) and to control N compounds peaks. Also, a mix of plant meals can be pelletized ("green-pellet") and applied into ponds [38]; or low protein diets containing high C:N ratio can also be carried out [16,33]. The carbon source serves as a substrate for operating BFT systems and production of microbial protein cells [36]. There are many considerations for its selection such as costs, local availability, biodegradability and efficiency of bacteria assimilation. In Table 1 is summarized some studies with different species and carbon source applied in BFT system.

Carbon source	Culture specie	Reference
Acetate	*Macrobrachium rosenbergii*	[39]
Cassava meal	*Penaeus monodon*	[40]
Cellulose	Tilapia	[12]
Corn flour	Hybrid bass and hybrid tilapia	[41, 42]
Dextrose	*Litopenaeus vannamei*	[43]
Glycerol and Glycerol+*Bacillus*	*M. rosenbergii*	[39]
Glucose	*M. rosenbergii*	[39]
Molasses	*L. vannamei and P. monodon*	[9, 29, 44]
Sorghum meal	Tilapia	[12]
Tapioca	*L. vannamei* and *M. rosenbergii*	[31, 45]
Wheat flour	Tilapia (*O. niloticus*)	[33]
Wheat bran + molasses	*Farfantepenaeus brasiensis, F. paulensis and F. duorarum*	[37, 46, 47]
Starch	Tilapia *O. niloticus x O. aureus* and tilapia (Mozambique)	[7, 14]

Table 1. Different carbon sources applied on BFT system (Source: adapted from [36])

Not all species are candidates to BFT. Some characteristics seems to be necessary to achieve a better growth performance such as resistance to high density, tolerance to intermediate levels of dissolved oxygen (~3-6 mg/L), settling solids in water (~10 with a maximum of 15 mL/L of "biofloc volume", measured in Imhoff cones) [38] and N-compounds, presence of filtering apparatus (i.e. tilapia), omnivorous habits and/or digestive system adaptable to better assimilate the microbial particles.

4. Applications in aquaculture

4.1. Nursery and grow-out

Nursery phase is defined as an intermediate step between hatchery-reared early postlarvae and grow-out phase [48]. Such phase presents several benefits such as optimization of farm land, increase in survival and enhanced growth performance in grow-out ponds [49-51]. BFT has been applied successfully in nursery phase in different shrimp species such as *L. vannamei* [44, 48], *P. monodon* [51], *F. paulensis* [15, 46], *F. brasiliensis* [37, 52] and *F. setiferus* [34]. The primary advantage observed is related to a better nutrition by continuous consumption of biofloc, which might positively influence grow-out performance *a posteriori* [53], but was not always the case [54]. In addition, optimization of farm facilities provided by the high stocking densities in BFT nursery phase seems to be an important advantage to achieve profitability in small farms, mainly in cold regions or when farmers are operating indoor facilities.

In [46] was observed that presence of bioflocs resulted in increases of 50% in weight and almost 80% in final biomass in *F. paulensis* early postlarval stage when compared to conventional clear-water system. This trend was observed even when postlarvae were not fed with a commercial feed (biofloc without commercial feed). In *L. vannamei* nursery in BFT conditions, references [48] and [55] reported survival rates ranging from 55.9% to 100% and 97% and 100%, respectively. In [51] was demonstrated that the addition of substrates in BFT systems increased growth and further enhanced production, while also contributing to more favorable water quality conditions. According to the same study, growth and survival was not affected by stocking density (2500 *vs* 5000 PL/m^2), therefore greater production outputs were achieved at the higher density. Furthermore, in [37] was found that *F. brasiliensis* postlarvae grow similarly with or without pelletized feed in biofloc conditions during 30-d of nursery phase, which was 40% more than conventional clear-water continuous exchange system.

In grow-out, BFT has been also shown nutritional and zootechnical benefits. In [9] was estimated that more than 29% of the daily food intake of *L. vannamei* consisted of microbial flocs, decreasing FCR and reducing costs in feed. The reference [10] showed that juveniles of *L. vannamei* fed with 35% CP pelletized feed grew significantly better in biofloc conditions as compared to clear-water conditions. In [28] was showed that controlling the concentration of particles in super-intensive shrimp culture systems can significantly improve shrimp production and water quality. Also, the same authors demonstrated that environmentally friendly plant-based diet can produce results comparable to a fish-based feed in BFT conditions. In [56] was evaluated the stocking density in a 120d of *L. vannamei* BFT culture, reporting consistent survival of 92, 81 and 75% with 150, 300 and 450 shrimp/m^2, respectively. Moreover, the study [57] performed in a heterotrophic-based condition detected no significant difference in FCR when feeding *L. vannamei* 30% and 45% CP diets and 39% and 43% CP diets, respectively. With these results in mind, floc biomass might provide a complete source of cellular nutrition as well as various bioactive compounds even at high density. It is not known exactly how microbial flocs enhance growth. Growth might be enhanced by continuous consumption of "native protein", protein source without previous treatment [18], which could possess a "growth factor" similar to the one investigated in squid [58]. Is well known that protein, peptides and aminoacids participate fully in synthesis of new membranes, somatic growth and immune function and biofloc can potentially provide such ingredients.

For fish and other species, BFT also has been demonstrated encouraged results. Intensive BFT *Oreochromis niloticus* tilapia culture could produce an equivalent of 155 ton/ha/crop [11]. Besides high yields, decrease of FCR and decreased of protein content in diets have also been observed. In [30] was estimated that feed utilization by tilapia is higher in BFT with a ration 20% less than conventional water exchange system. Studying the effect of BFT in juveniles tilapia, the reference [33] showed no difference in fish growth/production between 35% and 24% CP fed tanks under BFT, but both were higher than clear-water control without biofloc with 35% CP. Moreover, in [7] was investigated the effectiveness of BFT for maintaining good water quality in over-wintering ponds for tilapia. The authors concluded that BFT emerge as an alternative to overcome over-wintering problems, particularly mass mortality of fish due to low temperatures. In the study [14] was observed that biofloc consumed by fish (tilapia) may represent a very significant feed source, constituting about 50% of the regular feed ration of fish (assuming daily feeding of 2% body weight).

In *M. rosenbergii* larviculture was evaluated the effect of different carbon sources in a BFT culture conditions [39]. The authors found that using glucose or a combination of glycerol plus *Bacillus* as a carbon source in bioreactors led to higher biofloc protein content, higher n-6 fatty acids, which resulted in improved survival rates. In a study with a Brazilian endemic tropical fish species tambaqui (*Colossoma macropomum*) was observed that BFT did not improve fish growth/production as compared to clear-water conditions [59], although some water quality problems in such study remained unsolved (i.e. turbidity and nitrite). The authors showed no differences in 44% CP fed tanks under BFT and clear-water conditions, as well as 28% CP BFT. Certainly further research is needed to clarify the effect of BFT in *Colossoma macropomum*. On the other hand, *Piaractus brachypomus* or pirapitinga seems to be a candidate species to BFT [60].

4.2. Breeding

The BFT has been successfully applied for grow-out, but little is known about biofloc benefits on breeding. For example, in the shrimp industry with the global spread of viruses, the use of closed-life cycle broodstock appeared as a priority to guarantee biosecurity, avoiding vertical transmissions. Moreover, such industry places a considerable interest on penaeid breeding program, often performed in closed facilities, controlling the production plan through successive generations. These programs were frequently associated with large animals, disease resistance as well as the enhancement of reproductive performance. However, nutritional problems remain unresolved [61] and alternatives should be evaluated.

As an alternative for continuous *in situ* nutrition during the whole life-cycle, breeders raised in BFT limited or zero water exchange system are nutritional benefited by the natural productivity (biofloc) available 24 hours per day. Biofloc in a form of rich-lipid-protein source could be utilized for first stages of broodstock's gonads formation and ovary development. Furthermore, production of broodstock in BFT could be located in small areas close to hatchery facilities, preventing spread of diseases caused by shrimp transportation.

In conventional systems breeders used to be produced in large ponds at low density. However, risks associate with accumulation of organic matter, cyanobacteria blooms and fluctuations of some water quality parameters (such as temperature, DO, pH and N-compounds) remains high and could affect the shrimp health in outdoor facilities. Once the system is stable (sufficient particulate microbiota biomass measured in Imhoff cones), BFT provides stabilized parameters of water quality when performed in indoor facilities such as greenhouses, guaranteeing shrimp health.

According to studies performed with the blue shrimp *L. stylirostris* [18] and the pink shrimp *F. duorarum* broodstock [62], BFT could enhance spawning performance as compared to the conventional pond and tank-reared system, respectively (i.e. high number of eggs per spawn and high spawning activity; Fig 4). Such superior performance might be caused by better control of water quality parameters and continuous availability of food (biofloc) in a form of fatty acids protected against oxidation, vitamins, phospholipids and highly diverse "native protein", rather than conventional systems which "young" breeders are often limited to pelletized feed. These nutrients are required to early gonad formation in young breeders and subsequent ovary development. The continuous availability of nutrients could promote high nutrient storage in hepatopancreas, transferred to hemolymph and directed to ovary, resulting in a better sexual tissue formation and reproduction activity [18].

Regarding to shrimp broodstock management, one of the most important management procedures is related to control of solids and stocking density. High levels of solids negatively affect shrimp health, particularly with shrimp weight higher than 15g [47]. Settling solids or "biofloc volume" should be managed below than 15mL/L (measured in Imhoff cones) [38, 47]. Excess of particulate organic matter covered breeder's gills and could limit oxygen exchange, might resulting in mortalities.

Stocking density has to be carefully managed, mainly in sub-adult/adult phase (i.e. >15g). High density or high biomass will lead to an increase in organic matter, TSS levels and N-compounds in tanks or in ponds [63]. Moreover, physical body damages are prevented at low density, improving breeder's health. For review, a suggested stocking density is well described in [64].

For fish, no literature is available regarding BFT and application in breeders. The same trend observed in penaeid shrimp might be observed in fish. The continuous consumption of diverse microbiota (biofloc) should improve nutrients transfer, gonad formation and reproduction performance in fish. Lipid is a well-known nutrient that plays a key role in reproduction of aquatic species. In tilapia, breeders fed with crude palm oil based-feed (n-6 fatty acid rich source) presented high concentration of acid arachidonic or "ARA" (C20:4 n-6) in gonads, eggs and larvae of tilapia as compared to fish oil or linseed oil-based feeds [65]. As a result, better reproductive performance was observed in terms of higher total number of eggs per fish, larger gonad sizes, shorter latency period, inter spawning interval and higher spawning frequency. ARA is an essential fatty acid crucial in reproduction, acting as hormone precursor [66]. In the study [33] was found high ARA content in biofloc harvested in tilapia culture freshwater tanks. Bioflocs in freshwater bioreactors contained high ARA content using glucose and glycerol as a carbon source [67]. These findings suggested that biofloc (according its nutritional profile, for review see section 5.0) might positively influence the reproductive performance in fish, supplying nutrients for gonad development, possibly also enhancing larval quality *a posteriori*. BFT in tilapia broodstock could be an effective method to increase tilapia fry production and further research is need in this field.

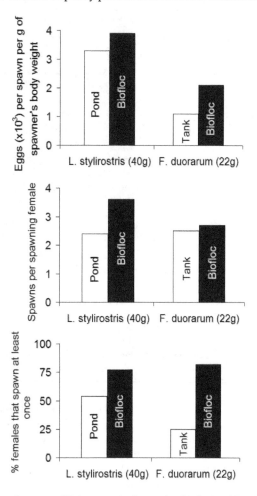

Figure 4. Spawning performance of *F. duorarum* (tank-reared vs biofloc) and *L. stylirostris* (pond-reared vs biofloc) performed in 45 and 30 days after ablation, respectively. Mean weights in parenthesis (more details in [18] and [62]).

4.3. The "natural probiotic" effect of biofloc

Biofloc can be a novel strategy for disease management in contrast to conventional approaches such as antibiotic, antifungal, probiotic and prebiotic application. The "natural probiotic" effect in BFT could act internally and/or externally against, i.e., to *Vibrio sp.* and ectoparasites, respectively. This effect is promoted by large groups of microorganisms, but mainly bacteria that is considered the first trophic level in the system.

Internally, bacteria and its synthesized compounds could act similar to organic acids and might be effective bio-control agents, also given beneficial host's microbial balance in the gut [68]. The regular addition of carbon in the water is known to select for polyhydroxyalkanoates (PHA) accumulating bacteria and other groups of bacteria that synthesize PHA granules. The microbial storage product poly-ß-hydroxybutyrate (PHB), a biodegradable polymer belonging to the polyesters class, is only one compound of a whole family of polyhydroxyalkanoates. PHB is produced by a widely variety of microorganisms such as *Bacillus sp., Alcaligenes sp., Pseudomonas sp.* from soluble organic carbon and is also involved in bacterial carbon metabolism and energy storage [68]. This polymer could comprise ~80% of the bacteria's cell dry matter and up to 16% on biofloc dry weight [69]. Different carbon sources or structures of carbon substrate will result in varying types of PHA [69].

Such granules are synthesized under conditions of physiological and nutrient stress, i.e., when an essential nutrient like nitrogen is limited in the presence of an excess carbon source [68]. When these polymers are degraded in the gut, they could have antibacterial activity similar to short chain fatty acids (SCFAs) or organic acids. The breakdown of PHA inside the gastrointestinal tract can be carried out via chemical and enzymatic hydrolysis [70].

Chemical hydrolysis can be carried out by treating the polymers with, i.e., NaOH, in which could significantly accelerate its digestibility [70]. On the other hand, enzyme hydrolysis is generally carried out by extracellular depolymerases activities which are widely distributed among bacteria and fungi, acting as a preventive or curative protector against *Vibrio sp.* infections and stimulate growth and survival of shrimp and fish larvae [69].

The working mechanism of PHAs with respect to their antibacterial activity is not well understood [68]. As they could act similarly to SCFA, some studies speculated the working mechanism by (i) reduction of pH, in which antibacterial activity increases with decreasing pH value [71]; (ii) inhibiting the growth of pathogenic bacteria by interference on cell membrane structure and membrane permeability, as well as instability of internal protons balance, lowering ATP and depletion of cellular energy [72]; and (iii) down-regulate virulence factor expression and positively influence the gut health of animals [73]. Further research is need to maximizing PHA content in bioflocs applied, i.e., for fish/shrimp feed, characterizing and analyzing their bio-control efficacy in different host-microbe systems [68].

Externally, the working mechanism of biofloc microorganisms against pathogens seems to be by competition of space, substrate and nutrients. Some essentials nutrients such as nitrogen are required by both groups (i.e. heterotrophic bacteria *vs Vibrio sp.*) limiting their growth. Inhibiting compounds excreted by BFT microorganisms, light intensity and type of carbon source also could reduce pathogens growth. Unfortunately, limited information is available on this field. In a study with fish fingerlings [74] was reported that tilapia (initial weight 0.98 ± 0.1g) reared under BFT limited water-exchange condition (FLOC) presented less ectoparasites in gills and ectoderm's mucous as compared to conventional water-exchange system (CW) after 60 days (Fig 5).

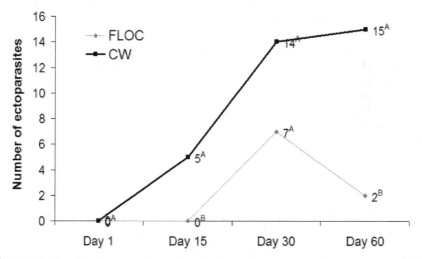

Figure 5. Number of total ectoparasites in gills and ectoderm's mucous of fry tilapia reared under BFT limited water-exchange condition (FLOC) and conventional water-exchange system (CW) after 60 days (more details in [74])

4.4. Aquaponics

Aquaponics is a sustainable food production system that combines a traditional aquaculture with hydroponics in a symbiotic environment. The water is efficiently recirculated and reused for maximum benefits through natural biological filtration and recirculation. The waste that is excreted by aquatic species or uneaten feed is naturally converted into nitrate and other beneficial nutrients in the water. Those nutrients are then absorbed by the vegetables and fruits in a "natural fertilization way".

Aquaculture species including fish, crayfish, freshwater prawns or shrimp are usually reared in tanks and the water directed into separated race-ways of hydroponics vegetables. A worldwide well-known aquaponics system was successfully developed by University of Virgin Islands (Fig 6). Typical plants raised in aquaponics include lettuce, chard, tomato, fruits such as passion fruit, strawberry, water melon, etc.; and a large variety of spices. Size of aquaculture tanks varies according aquatic species/vegetables demand and usual shapes includes round, square or rectangular tanks.

(Source: UVI website www.uvi.edu)

Figure 6. Aquaponics system at University of Virgin Islands

Nowadays, BFT have been successfully applied in aquaponics. The presence of rich-biota (microorganisms of biofloc) and a variety of nutrients such as micro and macronutrients originated from un-eaten or non-digested feed seems to contribute in plant nutrition. A well-known example of biofloc and aquaponics interaction was also developed by UVI. However, the application of BFT in aquaponics needs particular attention, mainly on management of solid levels in water (for review, see [28]). High concentration of solids may cause excessive adhesion of microorganism on plants roots (biofilm), causing its damage, lowering oxygenation and poor growth. Filtering and settling devices are often needed (Fig 7).

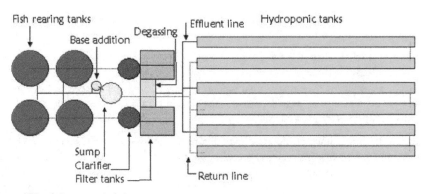

(Source: UVI website www.uvi.edu)

Figure 7. Scheme of worldwide well-known UVI Aquaponics System

5. Microbial biomass application in animal food industry

The cost of diets in several animal cultures is predominantly due to the cost of protein component [75]. In the case of aquaculture, its massive expansion in the last decades has begun to face some important limitations like increasing prices of fishmeal, a raw material prime component of aquaculture diets. However, pressure caused in natural stocks (over-fishing) has depleted fishmeal production and, as a consequence, continuous increase in prices has been observed [76]. Moreover, growth of aquafeed industry (driven by an increase in fish/shrimp demand as the global population continues to growth), the competition with other animal cultures (such as swine and poultry) and differences in fishmeal quality also collaborated with increase in prices of fishmeal. The quality attributed to fishmeal includes high palatability, high content of digestible protein, highly unsaturated fatty acids (HUFA) and minerals.

In this context, alternatives should be evaluated opposing this non-optimism scenario. Aquaculture industry needs to investigate alternative source of proteins to replace less sustainable ones. Candidates of protein sources might have good digestibility, palatability, energy content, low ash content and present a well-balanced essential amino acids profile (EAA) [77].

In the past years, BFT has been emerged not only as promising alternative to grow-out system, but also as a method to obtain protein for compounds diets originated from its diverse microbiota. Collected in tanks/ponds [46, 62] or produced in bioreactors [17, 39, 67] biofloc (Fig 8) is a raw material to produce "biofloc meal". In bioreactors, biofloc production can clean up effluent waters from aquaculture facilities, converting dissolved nutrients into single-cell protein [78]. Usually, two types of bioreactors have been employed: sequencing batch reactors (SBRs) and membrane batch reactors (MBRs), both controlling ammonia, nitrite and suspended solids with great efficacy (for review of bioreactors and its employ, see Kuhn et al 2012). Moreover, excess of solids removed from culture tanks or ponds and/or concentrated into solid removal devices [28] could also be a recyclable source of biofloc for biofloc meal production. This sustainable approach of protein source is getting more attention in the aquaculture industry. The microbial particles can provide important nutrients such as protein [33, 46], lipids [10, 37], aminoacids [80] and fatty acids [33, 67, 81].

Biofloc meal (also called "single-celled" protein), added to compounded feed is currently focus of intensive research in nutrition fields [17, 78]. However, to produce this protein ingredient some processes are required such as drying, milling and storage. In this context, nutritional characteristics could be affected (by i.e. temperature during drying), which the "native" properties could be altered.

Nutritional composition of biofloc differs according to environmental condition, carbon source applied, TSS level, salinity, stocking density, light intensity, phytoplankton and bacteria communities and ratio, etc. Regarding to age of bioflocs, in "young" biofloc heterotrophic bacteria is mainly presented as compared to "old" biofloc dominated by fungi [79]. In biofloc particles, protein, lipid and ash content could vary substantially (12 to 49, 0.5 to 12.5 and 13 to 46%, respectively; Table 2). The same trend occurs with fatty acids (FA) profile. Essential FA such as linoleic acid (C18:2 n-6 or LA), linolenic acid (C18:3 n-3 or ALA), arachidonic acid (C20:4 n-6 or ARA), eicosapentanoic acid (C20:5 n-3 or EPA) and docosahexaenoic acid (C22:6 n-3 or DHA), as well as sum of n-3 and sum of n-6 differ considerably between 1.5 to 28.2, 0.04 to 3.3, 0.06 to 3.55, 0.05 to 0.5, 0.05 to 0.77, 0.4 to 4.4 and 2.0 to 27.0% of total FA. Type of carbon source, freshwater or marine water and production of biofloc biomass (in bioreactors or culture tanks) definitely influence the FA profile (Table 3 and 4). Vitamin and amino acids profile from biofloc produced in large-scale commercial bioreactors [82] in given in Table 5.

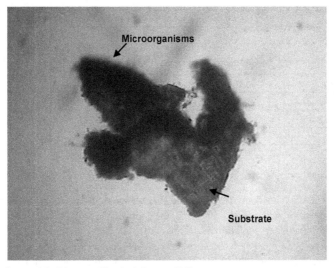

Figure 8. Biofloc particle (10x magnification) (Source: [54])

Information is still scarce about how microorganisms profile and its nutritional composition could impact animal growth. However, is already known that microorganisms in biofloc might partially replace protein content in shrimp diets, although were not always the case [10, 88]. Recent studies determined how reducing the protein content of diet would affect growth performance of shrimp reared in biofloc conditions. In the study [15] was found that at least 10% of protein content in pelletized feed can be reduced when *F. paulensis* postlarvae are raised in BFT conditions. In [89] was observed that shrimp fed with less than 25% crude protein under biofloc conditions performed similarly to shrimp raised under regular clear-water intensive culture with a 37%-protein diet. The biofloc system also delivered more consistent survival rates, especially at higher density. A low-protein biofloc meal-based pellet (25% CP) was evaluated as a replacement of conventional high-protein fishmeal diet (40% CP) for *L. vannamei* in a relatively low temperature (25°C) under biofloc conditions [35]. The results showed that is possible to replace 1/3 part of a conventional diet by alternative low-protein biofloc meal pellet without interfering survival and shrimp performance.

Crude protein (%)	Carbohydrates (%)	Lipids (%)	Crude fiber (%)	Ash (%)	Reference
43.0	-	12.5	-	26.5	[27]
31.2	-	2.6	-	28.2	[83]
12.0 - 42.0	-	2.0 - 8.0	-	22.0 - 46.0	[84]
31.1	23.6	0.5	-	44.8	[10]
26.0 - 41.9	-	1.2 - 2.3	-	18.3 - 40.7	[80]
30.4	-	1.9	12.4*	38.9	[85]
49.0	36.4	1.13	12.6	13.4	[17]
38.8	25.3	<0.1	16.2	24.7	[78]
28.8 - 43.1	-	2.1 - 3.6	8.7 - 10.4	22.1 - 42.9	[86]
30.4	29.1	0.5	0.8	39.2	[37]
18.2-29.3	22.8-29.9	0.4-0.7	1.5-3.5	43.7-51.8	[47]
18.4-26.3	20.2-35.7	0.3-0.7	2.1-3.4	34.5-41.5	[87]
28.0-30.4	18.1-22.7	0.5-0.6	3.1-3.2	35.8-39.6	[62]

*Lignin+cellulose

Table 2. Proximate analysis of biofloc particles in different studies.

Also, recent studies have been demonstrated that fishmeal in shrimp diets can be partially replaced by other protein sources under biofloc conditions or by biofloc meal. In [90] was evaluated two fishmeal replacement levels (40 and 100% of replacement) by other ingredients (soyabean meal and viscera meals) in diets for *Litopenaeus vannamei* reared in a biofloc system. The authors observed that fishmeal can be replaced in a level of 40.0% without interfering on growth performance and water quality. On the other hand, incorporating treated solids (microbial flocs) generated from tilapia effluent into shrimp feed, [91] demonstrated that shrimp performance was significantly increased as compared to untreated solids (settling basins of tilapia culture units). In [92] a trial performed in clear-water conditions detected that fishmeal can be completely replaced with soy protein concentrate and biofloc meal (obtained from super-intensive shrimp farm effluent) in 38% CP diets without adverse effects on *L. vannamei* performance. Moreover, [17] observed that biofloc produced in SBRs bioreactors using tilapia effluent and sugar as a growth media could offer an alternative protein source to shrimp feeds. Microbial floc-based diets significantly outperformed control fishmeal-based diets in terms of weight gain per week with no differences in survival.

Regarding to biofloc meal production, one bottleneck seems to be the large amount of wet biofloc biomass required to produce 1kg of dry biofloc meal. Estimative indicates that biofloc plug in 1L settling cones contained only 1.4% of dry matter [14]. The reference [17] indicated that 1 kg of microbial floc could be produced per 1.49 kg of sucrose in bioreactors. Certainly more research is needed on this field. On the other hand, other applications of biofloc meal in animal industry should be evaluated, mainly considering its nutritional profile and relatively low costs as compared to other protein sources (i.e. fishmeal) [17]. In aquaculture, biofloc meal could be included into broodstock pelletized feed, prior or after eyestalk ablation. Further research is encouraged in this field.

Fatty Acid	% of total fatty acid							
C14:0	0.10	0.60	0.80	0.45	1.43	0.69	0.61	0.43
C15:0	0.15	0.25	0.25	0.30	0.31	0.31	0.17	0.26
C16:0	2.2	17.0	26.0	15.0	6.06	8.01	6.34	8.86
C16:1	4.0	3.7	3.0	5.0	6.61	2.61	1.61	1.54
C17:0	0.05	0.4	0.5	0.2	0.20	0.23	0.14	0.68
C18:0	0.5	4.0	7.1	6.0	2.37	4.82	3.94	6.27
C18:1 n-7	1.5	3.0	1.9	2.7	3.96	1.72	2.71	4.19
C18:1 n-9	1.8	19.0	30.0	18.0	3.34	7.26	8.12	12.05
C18:2 n-6 (LA)	5.0	19.0	28.2	11.0	1.91	17.24	11.95	21.87
C18:3 n-3 (ALA)	0.04	0.5	0.45	2.0	0.23	0.99	0.20	0.21
C20:0	-	0.10	0.20	0.20	0.06	0.34	0.33	0.49
C20:1 n-9	0.05	0.10	0.15	0.10	0.25	0.20	0.06	0.02
C20:3 n-6	0.15	0.10	0.06	0.07	0.55	0.36	0.15	0.04
C20:4 n-6 (ARA)	0.7	0.3	0.15	0.20	0.77	0.87	0.17	0.06
C20:5 n-3 (EPA)	0.10	0.11	0.05	0.25	0.15	0.15	0.19	0.12
C22:6 n-3 (DHA)	0.05	-	0.07	0.05	0.18	0.06	0.18	0.10
$\sum$ Saturated	22.08	22.99	35.35	22.45	10.76	14.85	11.53	16.99
$\sum$ Monounsaturated	8.16	26.22	35.45	27.15	16.51	14.21	12.5	17.8
$\sum$ n-3	0.4	0.6	0.7	0.65	1.04	2.02	0.60	0.43
$\sum$ n-6	7.0	20.0	27.0	12.0	4.03	19.03	12.27	21.97
Type of water	freshwater	freshwater	freshwater	freshwater	freshwater	freshwater	marine	marine
Carbon source	Acetate	Glycerol	(Glycerol+ Bacillus)	Glucose	Glucose	Glycerol	Glucose	Glycerol
Collection	bioreactors	bioreactors	bioreactors	bioreactors	bioreactors	bioreactors	bioreactors	bioreactors
Reference	[39]				[67]			

Table 3. Fatty acid profile of biofloc (produced in experimental bioreactors) using different carbon source in marine water and freshwater

Fatty Acid	% of total fatty acid		
C14:0	2.02-2.48	13.8-16.1	5.4-6.2
C15:0	0.70-0.77	1.1-1.5	1.1-1.3
C16:0	17.88-19.10	45.4-53.5	48.7-49.3
C16:1	7.15-7.74	9.9-15.3	16.5-21.6
C17:0	-	0.7	0.9-1.0
C18:0	6.24-7.27	3.4-3.5	3.7-4.5
C18:1 n-7	11.05-11.28	-	-
C18:1 n-9	8.51-10.08	8.8-9.2	7.7-10.8
C18:2 n-6 9 (LA)	15.38-16.68	1.5-2.5	2.2-2.6
C18:3 n-3 (ALA)	0.65-0.73	2.0-2.3	2.2-3.3
C20:0	0.87-1.44	0.2-0.4	0.4
C20:1 n-9	0.74-0.80	0.3-0.4	0.5
C20:3 n-6	0.40-0.46	0.2	0.2
C20:4 n-6 (ARA)	3.11-3.55	0.3-0.4	0.3-0.4
C20:5 n-3 (EPA)	0.39-0.46	0.3-0.5	0.5
C22:6 n-3 (DHA)	0.74-0.77	0.2-0.4	0.3-0.4
$\sum$ Saturated	30.2-34.92	67.6-73.0	61.5-61.9
$\sum$ Monounsaturated	28.10-29-38	19.7-25.0	28.3-30.5
$\sum$ n-3	1.38-1.91	2.8-3.4	3.2-4.4
$\sum$ n-6	23.5-25.81	2.0-3.0	2.7-3.1
Type of water	freshwater	marine	marine
Carbon source	Wheat flour	molasses	molasses
Collection	Tilapia tanks	shrimp tanks	shrimp tanks
Reference	[33]	[87]	[62]

Table 4. Fatty acid profile of biofloc (collected in tanks) using different carbon source in marine water and freshwater

Amino Acids	As Fed (%)
Alanine	3.82
Arginine	3.60
Aspartic acid	6.36
Glutamic acid	8.04
Glycine	2.81
Histidine	1.46
Isoleucine	3.38
Leucine	5.06
Lysine	4.34
Methionine	1.41
Cysteine	0.55
Phenylalanine	3.29
Proline	2.77
Serine	2.82
Taurine	0.25
Threonine	3.11
Tryptophan	0.98
Tyrosine	2.83
Valine	3.52
Total	60.4
Vitamins	
Niacin	83.3 mg/kg
Thiamine B1	7.7 mg/kg
Riboflavin	39.0 mg/kg
Vitamin B12	12.0 mg/kg
Vitamin E	29.8 IU/kg

Table 5. Example of vitamin and amino acids profile from biofloc produced in large-scale commercial bioreactors [82].

6. Conclusions and perspectives of BFT

Biosecurity is a priority in aquaculture industry. For example, in shrimp farming, considerable impact of disease outbreaks during the past two decades greatly affected the operational management of shrimp farms worldwide [10]. Infected PLs and incoming water seem to be the main pathway for pathogen introduction. This scenario forced farmers to look for more biosecure culture practices to minimize the risk associated with exposure to pathogens [2]. Biofloc technology brings an obvious advantage of minimizing consumption and release of water, recycling *in situ* nutrients and organic matter. Furthermore, pathogens introduction is reduced, improving the farm biosecurity.

Biofloc technology will enable aquaculture grow towards an environmental friendly approach. Consumption of microorganisms in BFT reduces FCR and consequently costs in feed. Also, microbial community is able to rapidly utilize dissolved nitrogen leached from shrimp faeces and uneaten food and convert it into microbial protein. These qualities make minimal-exchange BFT system an alternative to extensive aquaculture. Microorganisms in biofloc might partially replace protein content in diets or decrease its dependence of fishmeal.

Related to biofloc meal and its perspectives, the study [17] detected initial estimates of cost for producing a metric ton of biofloc meal is approximately $400 to $1000. The same authors cited that global soymeal market varied approximately from $375 to $550/metric ton from January 2008 through May 2009. During the same time period, fishmeal varied approximately from $1000 to $1225, suggesting feasibility on replacement of either soybean and/or fish meal by biofloc meal. Moreover, generated from a process that cleans aquaculture effluents [17, 39] biofloc meal production avoids discharge of waste water and excessive damage to natural habitats [4]. This ingredient seems to be free of deleterious levels of mycotoxins, antinutritional factors and other constituents that limit its use in aquafeeds [79]. Large-scale production of biofloc meal for use in aquaculture could result in environmental benefits to marine and coastal ecosystems, as the need for wild fish as an aquafeed ingredient is reduced [79, 92].

Sensorial quality of BFT products is also an important issue. BFT may bring higher profit if fresh non-frozen shrimp/fish is sold to near-by market, mainly at inland locations. These advantages certainly should be more explored and niche markets achieved, contributing to social sustainability.

Author details

Maurício Emerenciano
Posgrado en Ciencias del Mar y Limnología, Universidad Nacional Autónoma de México (UNAM), Unidad Multidisciplinaria de Docencia e Investigación (UMDI), UNAM, Sisal, Yucatán, Mexico Santa Catarina State University, Centro de Educação Superior da Região Sul (CERES), Laguna, Santa Catarina, Brazil

Gabriela Gaxiola
UMDI, Facultad de Ciencias, UNAM, Sisal, Yucatán, México

Gerard Cuzon
Ifremer (Institut Français de Recherche pour l'Exploitation de la Mer) Taravao, Tahiti, French Polynesia

Acknowledgement

The authors would like to thank CONCYTEY (Consejo de Ciencia y Tecnología del Estado de Yucatán), Coordenação de Aperfeiçoamento de Pessoal de Nível Superior-CAPES,

Brazilian Ministry of Education (PhD grant number 4814061 provided to the primary author) and Consejo Nacional de Ciencia y Tecnología-CONACyT, México (grant 60824) for research support. The authors also would like to thank Wilson Wasielesky, Yoram Avnimelech and Manuel Valenzuela for photos courtesy and Miguel Arévalo, Maite Mascaró, Elsa Noreña, Santiago Capella, Adriana Paredes, Gabriela Palomino, Korynthia Aguiar, Moisés Cab, Nancy Aranda Cirerol, Concepción Burgos, Manuel Valenzuela and all staff of Programa Camarón-UMDI for their contribution towards researches performed at UMDI-UNAM cited in this chapter.

7. References

[1] Food and Agriculture Organization of the United Nations FAO (2008) Cultured aquaculture species information programme, *Penaeus vannamei* (Boone,1931). Food and Agriculture Organization of the United Nations. Available at http://www.fao.org

[2] Browdy CL, Bratvold D, Stokes AD, Mcintosh RP (2001) Perspectives on the application of closed shrimp culture systems. In: Jory ED, Browdy CL, editors. The new Wave, Proceedings of the Special Session on Sustainable Shrimp Culture, The World Aquaculture Society, Baton Rouge, LA, USA. pp.20–34.

[3] De Schryver P, Crab R, Defoirdt T, Boon N., Verstraete W (2008) The basics of bio-flocs technology: the added value for aquaculture. Aquaculture 277:125-137.

[4] Naylor RL, Goldburg RJ, Primavera JH, Kautsk N, Beveridge MCM, Clay J, Folke C, Lubchencoi J, Mooney H, Troell M (2000) Effect of aquaculture on world fish supplies. Nature 405:1017-1024.

[5] Bender J, Lee R, Sheppard M, Brinkley K, Philips P, Yeboah Y, Wah RC (2004) A waste effluent treatment system based on microbial mats for black sea bass *Centropristis striata* recycled water mariculture. Aquac Eng 31:73-82.

[6] Ray A (2012) Biofloc technology for super-intensive shrimp culture. In: Avnimelech Y, editor. Biofloc Technology - a practical guide book, 2nd ed., The World Aquaculture Society, Baton Rouge, Louisiana, USA. pp. 167-188.

[7] Crab R, Kochva M, Verstraete W, Avnimelech Y (2009) Bio-flocs technology application in over-wintering of tilapia. Aquac Eng 40:105-112.

[8] Burford MA, Thompson PJ, McIntosh RP, Bauman RH, Pearson DC (2003) Nutrient and microbial dynamics in high-intensity, zero-exchange shrimp ponds in Belize. Aquaculture 219:393–411.

[9] Burford MA, Thompson PJ, McIntosh RP, Bauman RH, Pearson DC (2004) The contribution of flocculated material to shrimp (*Litopenaeus vannamei*) nutrition in a high-intensity, zero-exchange system. Aquaculture 232:525–537.

[10] Wasielesky W.Jr, Atwood H, Stokes A, Browdy CL (2006) Effect of natural production in a zero exchange suspended microbial floc based super-intensive culture system for white shrimp *Litopenaeus vannamei*. Aquaculture 258:396–403.

[11] Rakocy JE, Bailey DS, Thoman ES, Shultz RC (2004) Intensive tank culture of tilapia with a suspended, bacterial based treatment process: new dimensions in farmed tilapia.

In: Bolivar R, Mair G, Fitzsimmons K, editors. Proceedings of the Sixth International Symposium on Tilapia in Aquaculture. pp. 584–596.

[12] Avnimelech Y, Mokady S, Schoroder GL (1989) Circulated ponds as efficient bioreactors for single cell protein production. Bamdigeh. 41:58–66.

[13] Hargreaves JA (2006) Photosynthetic suspended-growth systems in aquaculture. Aquac Eng 34:344-363.

[14] Avnimelech Y (2007) Feeding with microbial flocs by tilapia in minimal discharge bioflocs technology ponds. Aquaculture 264:140–147.

[15] Ballester ELC, Abreu PC, Cavalli RO, Emerenciano M, Abreu L, Wasielesky W (2010) Effect of practical diets with different protein levels on the performance of Farfantepenaeus paulensis juveniles nursed in a zero exchange suspended microbial flocs intensive system. Aquac Nut 16:163-172.

[16] Yoram Avnimelech (2012) Biofloc Technology - A Practical Guide Book, 2nd ed. The World Aquaculture Society, Baton Rouge, Louisiana, EUA. 272p.

[17] Kuhn DD, Boardman GD, Lawrence AL, Marsh L, Flick GJ (2009) Microbial flocs generated in bioreactors is a superior replacement ingredient for fishmeal or soybean meal in shrimp feed. Aquaculture 296:51–57.

[18] Emerenciano M, Cuzon G, Goguenheim J, Gaxiola G, Aquacop (2011) Floc contribution on spawning performance of blue shrimp *Litopenaeus stylirostris*. Aquac Res (published online first DOI: 10.1111/j.1365-2109.2011.03012.x)

[19] Aquacop (1975) Maturation and spawning in captivity of penaeid shrimp: *Penaeus merguiensis* de Man, *Penaeus japonicus* Bate, *Penaeus aztecus* Ives, *Metapenaeus ensis* de Haan and *Penaeus semisulcatus* de Haan. In: Avault W, Miller R, editors. Proceedings of the Sixth Annual Meeting World Mariculture Society, Lousiana State University, Baton Rouge, pp. 123–129.

[20] Sohier L (1986) Microbiologie appliquée à l'aquaculture marine intensive. pp. 119. Thèse Doctorat d'Etat, Université Aix-Marseille II Marseille, France.

[21] Cuzon G, Lawrence A, Gaxiola G, Rosas C, Guillaume J (2004) Nutrition of *Litopenaeus vannamei* reared in tanks or in ponds. Aquaculture 235:513–551.

[22] Rosenberry B (2010) Controlling pH in biofloc ponds. The shrimp news international. http://www.shrimpnews.com/FreeReportsFolder/phContronBioflocPonds.html. Acessed in 11/02/2012.

[23] Garen P, Aquacop (1993) Nuevos resultados en la cría intensiva de camarón *Penaeus vannamei* y *P. stylirostris*. In: Calderón JV, Sandoval VC, editors. Memorias del I Congresso Ecuatoriano de Acuicultura, Guayaquil, 18-23 Octubre 1992. Escuela Superior Politécnica del Litoral, Guayaquil, pp. 137–145.

[24] Rosenberry B (2007) Marvesta Shrimp Farms Delivers "Fresh" Shrimp to Restaurants. The shrimp news international. http://www.shrimpnews.com/FreeNewsFolder/FreeNewsBackIssues/2007BackIssues/FreeNewsApril200720.html. Acessed in 12/02/2012.

[25] Moriarty DJW (1997) The role of microorganisms in aquaculture ponds. Aquaculture 151:333–349.

[26] Ray AJ, Seaborn G, Leffler JW, Wilde SB, Lawson A, Browdy CL (2010) Characterization of microbial communities in minimal-exchange, intensive aquaculture systems and the effects of suspended solids management. Aquaculture 310:130–138.

[27] McIntosh D., Samocha T.M., Jones E.R., Lawrence A.L., McKee D.A., Horowitz S. & Horowitz A. (2000) The effect of a bacterial supplement on the high-density culturing of *Litopenaeus vannamei* with low-protein diet in outdoor tank system and no water exchange. Aquac Eng 21:215–227.

[28] Ray AJ, Lewis BL, Browdy CL, Leffler JW (2010) Suspended solids removal to improve shrimp (*Litopenaeus vannamei*) production and an evaluation of a plant-based feed in minimal-exchange, superintensive culture systems. Aquaculture 299:89-98.

[29] Panjaitan P (2004). Field and laboratory study of *Penaeus monodon* culture with zero water exchange and limited water exchange model using molasses as a carbon source. Ph.D. Thesis, Charles Darwin Univ., Darwin, NT, Australia.

[30] Avnimelech Y, Kochva M, Diab S (1994) Development of controlled intensive aquaculture systems with a limited water exchange and adjusted carbon to nitrogen ratio. Bamidgeh 46:119–131.

[31] Hari B, Kurup BM, Varghese JT, Schrama JW, Verdegem MCJ (2004) Effects of carbohydrate addition on production in extensive shrimp culture systems. Aquaculture 241:179–194.

[32] Schneider O, Sereti V, Eding EH, Verreth JAJ (2005) Analysis of nutrient flows in integrated intensive aquaculture systems. Aquac Eng 32:379–401.

[33] Azim ME, Little DC (2008) The biofloc technology (BFT) in indoor tanks: Water quality, biofloc composition, and growth and welfare of Nile tilapia (*Oreochromis niloticus*). Aquaculture 283:29–35.

[34] Emerenciano M, Vinatea L, Gálvez AG, Shuler A, Stokes A, Venero J, Haveman J, Richardson J, Thomas B, Leffler J (2009) Effect of two different diets fish meal based and "organic" plant based diets in *Litopenaeus setiferus* earlier post-larvae culture under biofloc, green-water and clear-water conditions. CD of abstracts of World Aquaculture Society Meeting 2009, Veracruz, México.

[35] Emerenciano M, Cuzon G, López-Aguiar K, Noreña-Barroso E, Máscaro M, Gaxiola G. (2011) Biofloc meal pellet and plant-based diet as an alternative nutrition for shrimp under limited water exchange systems. CD of abstracts of World Aquaculture Society Meeting 2011, Natal, RN, Brazil.

[36] Avnimelech Y (1999) Carbon and nitrogen ratio as a control element in aquaculture systems. Aquaculture 176: 227–235.

[37] Emerenciano M, Ballester ELC, Cavalli RO, Wasielesky W (2012) Biofloc technology application as a food source in a limited water exchange nursery system for pink shrimp *Farfantepenaeus brasiliensis* (Latreille, 1817). Aquac Res 43:447-457.

[38] Taw N (2010) Biofloc technology expanding at white shrimp farms. Global Advocate may/june, 24–26 (available in http://www.gaalliance.org/mag/May_June2010.pdf)

[39] Crab R, Chielens B, Wille M, Bossier P, Verstraete W (2010) The effect of different carbon sources on the nutritional value of bioflocs, a feed for *Macrobrachium rosenbergii* postlarvae. Aquac Res 41: 559-567.

[40] Avnimelech Y, Mokady S (1988) Protein biosynthesis in circulated ponds. In: Pullin RSV, Bhukaswan T, Tonguthai K, Maclean JL, editors. Proceedings of Second International Symposium on Tilapia in Aquaculture, Departament of Fisheries of Thailand and ICLARM, Manila, Philippines, pp. 301-309.

[41] Milstein A, Anvimelech Y, Zoran M, Joseph D (2001) Growth performance of hybrid bass and hybrid tilapia in conventional and active suspension intensive ponds. Bamidgeh 53:147-157.

[42] Asaduzzaman M, Rahman MM, Azim ME, Islam MA, Wahab MA, Verdegem MCJ, Verreth JAJ (2010). Effects of C/N ratio and substrate addition on natural food communities in freshwater prawn monoculture ponds. Aquaculture 306: 127-136.

[43] Suita SM (2009) O uso da Dextrose como fonte de carbono no desenvolvimento de bio-flocos e desempenho do camarão-branco (*Litopenaeus vannamei*) cultivado em sistema sem renovação de água. Tese de mestrado. Universidade Federal do Rio Grande, Rio Grande do Sul, Brasil.

[44] Samocha TM, Patnaik S, Speed M, Ali AM, Burger JM, Almeida RV, Ayub Z, Harisanto M, Horowitz A, Brock DL (2007) Use of molasses as carbon source in limited discharge nursery and grow-out systems for *Litopenaeus vannamei*. Aquac Eng 36:184-191.

[45] Asaduzzaman M, Wahab MA, Verdegem MCJ, Huque S, Salam MA, Azim ME (2008) C/N ratio control and substrate addition for periphyton development jointly enhance freshwater prawn *Macrobrachium rosenbergii* production in ponds. Aquaculture 280, 117-123.

[46] Emerenciano M, Ballester ELC, Cavalli RO, Wasielesky W (2011b) Effect of biofloc technology (BFT) on the early postlarval stage of pink shrimp *Farfantepenaeus paulensis*: growth performance, floc composition and salinity stress tolerance. Aquac Int 19:891-901.

[47] Emerenciano M, Cuzon G, Paredes A, Gaxiola G (2012) Biofloc technology applied to intensive broodstock farming of pink shrimp *Farfantepenaeus duorarum* (Part I): grow-out, water quality, microorganisms profile and proximate analysis of biofloc. Aquac Res (*submitted*)

[48] Mishra JK, Samocha TM, Patnaik S, Speed M, Gandy RL, Ali A (2008) Performance of an intensive nursery system for the Pacific white shrimp, *Litopenaeus vannamei*, under limited discharge condition. Aquac Eng 38:2-15.

[49] Apud FD, Primavera JH, Torres PL (1983) Farming of Prawns and Shrimps. Extension Manual (5), SEAFDEC Aquaculture Department, Iloilo, Philippines. 67pp.

[50] Sandifer PA, Stokes AD, Hopkins JS (1991) Further intensification of pond shrimp culture in South Carolina. In: Sandifer PA, editor. Shrimp Culture in North America and the Caribbean, World Aquaculture Society, Baton Rouge, LA, USA. pp.84–95.

[51] Arnold SJ, Coman FE, Jackson CJ, Groves SA (2009) High-intensity, zero water-exchange production of juvenile tiger shrimp, *Penaeus monodon*: An evaluation of artificial substrates and stocking density. Aquaculture 293:42-48.

[52] Souza DM, Suita SM, Leite FPL, Romano LA, Wasielesky W, Ballester ELC (2011). The use of probiotics during the nursery rearing of the pink shrimp *Farfantepenaeus brasiliensis* (Latreille, 1817) in a zero exchange system. Aquac Res (published online first doi:10.1111/j.1365-2109.2011.02992.x)

[53] AQUACOP, Le Moullac G, Damez D (1991) Modélisation de la résistance au chocs de salinité des postlarves de Penaeus vannamei. Aquat Living Resour 4:169–173

[54] Emerenciano MGC, Wasielesky W, Soares RB, Ballester EC, Cavalli RO, Izeppi EM (2007) Crescimento e sobrevivêcia do camarão-rosa *Farfantepenaeus paulensis* na fase de berçário em meio heterotrófico. Acta Scientiarum Biological Sciences 29, 1–7.

[55] Cohen J, Samocha TM, Fox JM, Gandy RL, Lawrence AL (2005) Characterization of water quality factors during intensive raceway production of juvenile L. *vannamei* using limited discharge and biosecure management tools. Aquac Eng 32:425–442.

[56] Krummenauer D, Peixoto S, Cavalli RO, Poersch LH, Wasielesky W (2011) Superintensive culture of white shrimp, *Litopenaeus vannamei*, in a biofloc technology system in southern brazil at different stocking densities. J. World Aquacult. Soc. 42:726-733.

[57] Samocha TM, Patnaik S, Gandy RL(2004) Heterotrophic intensification of pond shrimp production. Book of abstract of Fifth International Conference on Recirculating Aquaculture, 22– 25 July 2004, Roanoke, Virginia, USA.

[58] Cruz-Ricque LE, Guillaume J, Cuzon G (1987) Squid protein effect on growth of four penaeid shrimp. J. World Aquacult. Soc. 18: 209–217.

[59] Itani AL, Neto ETA, Silva SL, Araújo ML, Lima AF, Barbosa JM (2010) Efeito do sistema heterotrófico no crescimento do Tambaqui (Colossoma macropomum). X JORNADA DE ENSINO, PESQUISA E EXTENSÃO – JEPEX 2010 – UFRPE: Recife.

[60] Poleo G, Aranbarrio JV, Mendoza L, Romero O (2011) Cultivo de cachama blanca en altas densidades y en dos sistemas cerrados. Pesq. Agropec. Bras. 46: 429-437

[61] Wouters R, Lavens P, Nieto J, Sorgeloos P (2001a) Penaeid shrimp broodstock nutrition: an updated review on research and development. Aquaculture 202:1–21.

[62] Emerenciano M, Cuzon G, Arévalo M, Gaxiola G (2012) Biofloc technology applied to intensive broodstock farming of pink shrimp *Farfantepenaeus duorarum* (Part II): spawning performance, biochemical composition and fatty acid profile. Aquac Res (submitted)

[63] Vinatea L, Gálvez AO, Browdy CL, Stokes A, Venero J, Haveman J, Lewis BL, Lawson A, Shuler A, Leffler JW (2010) Photosynthesis, water respiration and growth

performance of *Litopenaeus vannamei* in a super-intensive raceway culture with zero water exchange: Interaction of water quality variables. Aquac Eng 42:17–24.

[64] Emerenciano M, Gaxiola G, Cuzon G (2012). Biofloc Technology applied to shrimp broodstock. In: Avnimelech Y, editor. Biofloc Technology - apractical guide book, 2nd ed., The World Aquaculture Society, Baton Rouge, Louisiana, USA. pp. 217-230.

[65] Ng W, Wang Y (2011) Inclusion of crude palm oil in the broodstock diets of female Nile tilapia, *Oreochromis niloticus*, resulted in enhanced reproductive performance compared to broodfish fed diets with added fish oil or linseed oil. Aquaculture 314:122–131.

[66] Glencross BD (2009) Exploring the nutritional demand for essential fatty acids by aquaculture species. Rev Aquac 1:71-124.

[67] Ekasari J, Crab R, Verstraete W (2010) Primary nutritional content of bio-flocs cultured with different organic carbon sources and salinity. HAYATI Journal of Biosciences 17:125-130.

[68] Sinha AK, Baruah K, Bossier P (2008) Horizon Scanning: the potential use of biofloc as an anti-infective strategy in aquaculture – an overview. Aquac Health Int 13:8-10.

[69] De Schryver P, Boon N, Verstraete W, Bossier P. (2012) The biology and biotechnology behind bioflocs. In: Avnimelech Y, editor. Biofloc Technology - a practical guide book, 2nd ed., The World Aquaculture Society, Baton Rouge, Louisiana, USA. pp. 217-230.

[70] Yu J, Plackett D, Chen LXL (2005). Kinetics and mechanism of the monomeric products from abiotic hydrolysis of poly [(R) -3-hydroxybutyrate] under acidic and alkaline conditions. Polym Degrad Stabil 89:289-299.

[71] Ricke SC (2003). Perspectives on the use of organic acids and shortchain fatty acids as antimicrobials. Poult Sci 82:632-639.

[72] Russel JB (1992). Another explanation for the toxicity of fermentationacids at low pH: anion accumulation versus uncoupling. J Appl Bacteriol 73:363-370.

[73] Teitelbaum JE, Walker WA (2002) Nutritional impact of pre and pro-biotics as protective gastrointestinal organisms. Annu Rev Nutr 22:107-138.

[74] Emerenciano M, Avnimelech Y, Gonzalez R, Leon ATD, Cuzon G, Gaxiola G (2009) Effect of bio-floc technology (BFT) in ectoparasite control in Nile tilapia Oreochromis niloticus culture. CD of Abstracts of World Aquaculture Society Meeting 2009, Veracruz, Veracruz, Mexico.

[75] Tacon AGJ (1990) Standard methods for the nutrition and feeding of farmed fish and shrimp, Washington DC, Argent Laboratories Press, 454 pp.

[76] Tacon AGJ (1997) Feeding tomorrow's fish: keys for sustainability. In: Tacon A, Basurco B, editors. Feeding tomorrow's fish. Proceedings of the workshop of the CIHEAM Network on Technology of Aquaculture in the Mediterranean (TECAM), jointly organized by CIHEAM, FAO and IEO Mazarron, Spain, 24-26 June 1996, CIHEAM, Apodo, Spain, pp. 11-33.

[77] Hardy RW (2010) Utilization of plant proteins in fish diets effects of global demand and supplies of fishmeal. Aquac Res 41:770-776.

[78] Kuhn DD, Lawrence AL, Boardman GD, Patnaik S, Marsh L, Flick GJ (2010) Evaluation of two types of bioflocs derived from biological treatment of fish effluent as feed ingredients for Pacific white shrimp, *Litopenaeus vannamei*. Aquaculture 303:28–33.

[79] Kuhn DD, Lawrence A (2012) Ex-situ biofloc technology. In: Avnimelech Y, editor. Biofloc Technology - a practical guide book, 2nd ed., The World Aquaculture Society, Baton Rouge, Louisiana, USA. pp. 217-230.

[80] Ju ZY, Forster I, Conquest L, Dominy W, Kuo WC, Horgen FD (2008) Determination of microbial community structures of shrimp floc cultures by biomarkers and analysis of floc amino acid profiles. Aquac Res 39:118-133.

[81] Izquierdo M, Forster I, Divakaran S, Conquest L, Decamp O, Tacon A (2006). Effect of green and clear water and lipid source on survival, growth and biochemical composition of Pacific white shrimp *Litopenaeus vannamei*. Aquac Nut 12:192–202.

[82] Logan AJ, Lawrence A, Dominy W, Tacon AGJ 2010. Single-cell proteins from food by-products provide protein in aquafeed. Global Advocate 13:56-57.

[83] Tacon AGJ, Cody JJ, Conquest LD, Divakaran S, Forster IP, Decamp OE (2002) Effect of culture system on the nutrition and growth performance of Pacific white shrimp *Litopenaeus vannamei* (Boone) fed different diets. Aquac Nut 8:121–137.

[84] Soares R, Jackson C, Coman F, Preston N (2004) Nutritional composition of flocculated material in experimental zero-exchange system for *Penaeus monodon*. In: Proceedings of Australian Aquaculture, 2004, WAS, Sydney p.89

[85] Ju ZY, Forster I, Conquest L, Dominy W (2008) Enhanced growth effects on shrimp (*Litopenaeus vannamei*) from inclusion of whole shrimp floc or floc fractions to a formulated diet. Aquac Nut 14:533–543.

[86] Maicá PF, Borba MR, Wasielesky W (2012) Effect of low salinity on microbial floc composition and performance of *Litopenaeus vannamei* (Boone) juveniles reared in a zero-water-exchange super-intensive system. Aquac Res 43:361–370.

[87] Emerenciano M, Cuzon G, Arévalo M, Miquelajauregui MM, Gaxiola G (2012) Effect of short-term fresh food supplementation on reproductive performance, biochemical composition and fatty acid profile of Litopenaeus vannamei (Boone) reared under biofloc conditions. Aquac Int (submitted).

[88] McIntosh D, Samocha TM, Jones ER, Lawrence AL, Horowitz S, Horowitz A (2001) Effects of two commercially available low protein diets (21% and 31%) on water sediment quality, and on the production of *Litopenaeus vannamei* in an outdoor tank system with limited water discharge. Aquac Eng 25:69–82.

[89] Nunes AJP, Castro LF, Sabry-Neto H (2010) Microbial flocs spare protein in white shrimp diets. Global Advocate 10:28-30.

[90] Scopel RB, Schveitzer R, Seiffert WQ, Pierri V, Arantes RF, Vieira FN, Vinatea LA (2011) Substituição da farinha de peixe em dietas para camarões marinhos cultivados em sistema bioflocos. Pesq Agropec Bras 46: 928-934

[91] Kuhn DD, Boardman GD, Craig SR, Flick GJ, Mclean E (2008) Use of microbial flocs generated from tilapia effluent as a nutritional supplement for shrimp, *Litopenaeus vannamei*, in recirculating aquaculture systems. J. World Aquacult. Soc. 39:72–82.

[92] Bauer W, Prentice-Hernandez C, Tesser MB, Wasielesky W, Poersch LHS (2012) Substitution of fishmeal with microbial floc meal and soy protein concentrate in diets for the pacific white shrimp Litopenaeus vannamei. Aquaculture 342-343; 112–116.

Chapter 8

Phytoplankton Biomass Impact on the Lake Water Quality

Ozden Fakioglu

Additional information is available at the end of the chapter

1. Introduction

Pyhtoplankton is a plant plankton which cannot move actively and changes location depending on the movement of water. Phytoplankton communities are widely spreaded from aquatic to terresial lands. Plankton form the first ring of food chain in aquatic environment effecting the efficiency of this environment. Daily, seasonally and yearly changes are important for calculating efficiency in aquatic fields. Phytoplankton composition is a trophic indication of the water mass. In addition, phytoplankton species are used as an indicator for determining the nutrient level which is the basis for preparing and monitoring the strategies of the lake management in the lakes. Using phytoplankton communities or other aquatic organisms for evaluating water quality is based on very ancient times. Saprobik and trophic inducator types are used in many researches [1-3]. In addition, various numerical indices have been developed [1-2]. However, none of them have been accepted extensively. This is caused by several reasons. Those reasons are:

a. Differences in phytoplankton groups and group concept
b. Dynamic properties of phytoplankton groups
c. Habitat diversity of freshwater ecosystems
d. Phytogeographical differences [4].

Phytoplankton communities are influenced by significant changes every year. The competitive environment known as seasonal succession has been changing [5]. If the conditions don't change, this process results in the choice of communities dominated by one or more species. Phytoplankton responds rapidly to the condition chages. Conditional changes will result the formation of high compositional diversity [6].

The first approaches to classify algal communities don't have wide range of application in determining water quality. Pankin's approaches to classify algal communities in 1941 and 1945 weren't generally accepted [4]. Reynold's [7-8] applied a classic phytosociological approach to phytoplankton data obtained from the lakes in northeast england and classified them into various communities. Sommer [5] found high similarities among the compositions of species and seasonal succession of Alpine lakes. Mason [9] reported that oligotrophic and eutrophic lakes have the communities of characteristic phytoplankton.

There are qualitative differences among phytoplankton communities in oligotrophic and eutrophic lakes. The compositon and the amoung of phytoplankton communities are affected by environmental conditions. For example, a numerical decrease is observed in *Anabeana* and *Aphanizomenon* species from heterosis blue-green algae found in mesotrophic lake layers with a decrease of nitrogen saturation in the lakes. In addition, there are differences among the environmental conditions preferred by *Diatoms, Dinoflagellate* or *Cosmarium, Pandorina* and *Gemellicystis* species, even though they can be found in the lakes with same level of nutrients [10].

In the works related with phytoplankton communities used for predicting the ecological structure of aquatic systems, it has been tried to develop functional groups by improving the systematic investigation. Furthermore, some indices was developed according to numerical and biovolume values of phytoplankton (Palmer Index (1969), Descy Index (1979), TDI Index (1995) etc.). HPLC pigment analysis is used for the diagnosis of phytoplankton species in recent years [11].

The methods we will mention for examination of plankton in aquatic environments are summarized by compiling researchers' methods and techniques. The main target of these suggested methods is to obtain values close to the actual volume and weight of plankton in freshwater and to calculate the volume and weight (biomass) of these organisms by this method.

Besides, the methods and techniques in this subject are sensitive, determining ecological parameters which are characterizing the aquatic environment are important for the researches. Sampling error in this review may cause errors far beyond the susceptibility of calculations. In addition, vertically and horizontally distributions of plankton may show big changes against the effects of wind and light. For this reason, evaluating various samples collected as verticaly and horizontaly causes to get more reliable results from each sample.

Many techniques have been developed depending on the number, volume and cell structures of fresh water phytoplankton. In this section, studies conducted by calculating biovolume of phytoplankton used for estimating ecological characteristics of freshwater ecosystems will be summarized.

2. Changes in biomass of phytoplankton in lakes

The composition and biomass of phytoplankton are very important parameters for understanding the structure and tropic level of aquatic systems. Phytoplankton cell size, carbon content and functional structure are investigated by many researchers. Phytoplankton communities can have cell size from a few microns to a few milimeters depending on the groups they belong to. Biovolume measurements are estimated by automatic or semi-automatic methods. For example, morphometric methods, holographic method etc. In addition to these, the commonly used one is geometric method [12]. In addition to these methods, the most common method for calculationg biomass is measuring chlorophyl a value. Chlorophyll a is an important photosynthetic pigment for plant organisms. Environmental factors affect the amount of ambient phytoplankton and chlorophyll a value changes depending on the amount of phytoplankton [13].

Coulter Counter method is based on measuring electrical conductivity among cells. Electric current flowing through the cells placed in physiological saline varies depending on the cell size. Thus, the cell size is determined.

Morphometric methods is used in determinining the quantitative properties of cells. With this method, cell density is calculated depending on cell wall, chloroplast, vacuole, depot material and cytoplasmic content [14].

Holographic scanning tecnology which is used in conjunction with one curved mirror to passively correct focal plane position errors and spot size changes caused by the wavelength instability of laser diodies [15].

Geometric method is based on estimating biomass of phytoplankton via geometric shapes and mathematical equations. This model was found by Kovada and Larrance. 20 geometric models developed by Hillebrand et al. [16] are used for calculating biomass of algae. Each model was designed depending on cell structure with the shapes like sphere, cone, triangle etc. This method was applied to phytoplankton species found in sea waters in China and 31 geometric models were developed in this study [12].

Phytoplankton communities show vertical changes from time to time. Chlorophyll a is a pigment used for estimating biomass of phytoplankton. Seasonal changes cause variation in chlorophyll a value. For this reason, water affects the production of column light transmittance, hence, the value of chlorophyll a [18]. In determining chlorophyll a, fiber glass filter papers used for filtering water samples are waited for 3-4 hours then they are decomposed and kept in 10 ml 90% acetone one night, centrifuged and optical density of the extract is made by reading from spectrophotometer with 630, 645 and 665 nm wavelength [17].

The first step for calculating phytoplankton biomass is to store and protect the phytoplankton samples.

3. Storing and protecting phytoplankton samples

Before collecting phytoplankton samples from the lake, turbidity and temperature of the water were measured by Secchi disk. Phytoplankton samples have to be stored in 100-150 ml glass or polyethylene containers with 2% Lugol's solution or 4% Buffere formalin solution [32].

4. Identification of phytoplankton

Water samples taken by phytoplankton scoop were identified according to world literature [19-29].

5. Phytoplankton caunting

Water samples put into hydrobios plankton counting chambers depending on phytoplankton density, after standing overnight by dropping lugol's solution, counting phytoplankton were made by using inverted microscope [30-31].

Following formula was used to calculate the number of phytoplankton [31]:

$$\text{Number of phytoplankton (piece/ml)} = \frac{C \times TA}{F \times A \times V}$$

Here;

C= The number of organisms found by counting (number),
TA= Bottom area of the cell count (mm²),
F= Counted field number (number),
A= Field of view of the microscope (mm²),
V= Volume of precipitated sample (ml).

6. Estimation of phytoplankton biomass

Phytoplankton analysis is possible by a simple Kolkwitz chamber. Except deposited plankton in sample bottle, liquid at the top pour a few millilitres and centrifuged and their volumes are measured in sedimentation tubes then they are transferred to Kolkwitz chamber for analysis. In this method, phytoplankton analysis is the first, semi-field analysis of Kolkwitz chamber is the second and the third one is the analysis of the various fields.

Biovolume was estimated in the measurement of phytoplankton biomass. Phytoplankton were emulated to the geometric shapes like sphere, cylinder and cone and necessary measurements were taken from the phytoplankton while counting [32]. Geometric shapes and calculations used for calculating biovolume was done according to the formulas (Table 1) stated by [12] and [16]. After calculating average volume of every species, total volume were calculated by multiplying with the number of species. Following formula was used to calculate total cell volume of phytoplankton [31];

$$HH = \sum_{i=1}^{n} (HNixSHi)$$

Here;

HH= Total biovolume of plankton (mm/l),
HNi= the number of organisms belongs to i. species /l,
SHi= Average cell volume of i. species.

Biovolume is calculated by assuming cell volume is equal to 1mg age weight/m³ algal biovolume for 1mm/ m³ [33].

Shape	Biovolume	Samples
	$V = \dfrac{\pi}{6} \cdot a^3$	Crucigeniella apiculata Gomphosphaeria sp. Anabeana sp.
	$V = \dfrac{\pi}{6} \cdot b^2 \cdot a$	Coelastrum microporum Actinastrum hantzschii Dinobryon divergens Cryptomonas sp. Pandorina sp.
	$V = \dfrac{\pi}{6} \cdot a \cdot b \cdot c$	Trachelomonas caudata Peridinium sp. Botryococcus braunii Cocconeis placentula Phacus tortus
	$V = \dfrac{\pi}{4} \cdot a^2 \cdot c$	Cyclotella sp. Mougeotia sp.

Table 1. Geometrical shapes and formulas for calculating biovolume (continued)

Shape	Biovolume	Samples

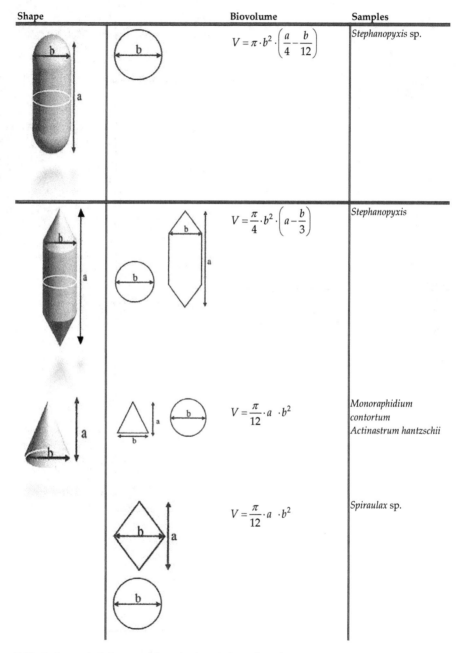

$$V = \pi \cdot b^2 \cdot \left(\frac{a}{4} - \frac{b}{12} \right)$$ *Stephanopyxis* sp.

$$V = \frac{\pi}{4} \cdot b^2 \cdot \left(a - \frac{b}{3} \right)$$ *Stephanopyxis*

$$V = \frac{\pi}{12} \cdot a \cdot b^2$$ *Monoraphidium contortum*
Actinastrum hantzschii

$$V = \frac{\pi}{12} \cdot a \cdot b^2$$ *Spiraulax* sp.

Table 1. Geometrical shapes and formulas for calculating biovolume (continued)

Shape		Biovolume	Samples
		$V = \dfrac{\pi}{4} \cdot b^2 \cdot a$	*Chroomonas* sp.
		$V = a \cdot b \cdot c$	*Asterionella* sp. *Synedra* sp. *Merismopedia* sp. *Epithemia zebra* var. *saxonica*
		$V = \dfrac{\pi}{4} \cdot a \cdot b \cdot c$	*Pediastrum* sp. *Navicula* sp.
		$V = \dfrac{\pi}{4} \cdot a \cdot b \cdot c$	*Cymatopleura* sp.
		$V = \dfrac{1}{2} \cdot a \cdot b \cdot c$	*Nitzschia* sp.
		$V = \dfrac{\pi}{4} \cdot a \cdot b \cdot c$	*Phaeodactylum* sp.

Table 1. Geometrical shapes and formulas for calculating biovolume (continued)

Shape		Biovolume	Samples
		$V \cong \dfrac{\pi}{4} \cdot a \cdot b \cdot c$ $V \cong \dfrac{\pi}{6} \cdot a \cdot b^2$	*Monorophidium* sp. *Eunotia* sp.
		$V = a \cdot c^2 \cdot a\sin\left(\dfrac{b}{2c}\right)$	*Cymbella* sp. *Amphora ovalis* *Epithemia* sp. *Rhopalodia gibba*
		$V = \dfrac{\sqrt{3}}{4} \cdot c \cdot a^2$	*Hydrosera* sp.
		$V = \dfrac{1}{6} \cdot a^2 \cdot c$	*Tetradinium*

Table 1. Geometrical shapes and formulas for calculating biovolume (continued)

Shape		Biovolume	Samples
		$$V \cong \frac{\pi}{4} \cdot a \cdot b \cdot c$$	*Tabellaria* sp.
		$$V \cong \frac{a \cdot b}{4} \cdot \left[a + \left(\frac{\pi}{4} - 1 \right) \cdot b \right] \cdot$$ $$a \sin \left(\frac{c}{2a} \right)$$	*Gomphonema constrictum*
		$$V = \frac{\pi}{3} \cdot (a_1 + a_2) \cdot b_1^2 + \frac{\pi}{4} \cdot$$ $$(a_2 + b_2) \cdot b_2^2 + \frac{\pi}{12} \cdot a_2 \cdot b_1 \cdot b_2$$	*Euglena* sp.

Table 1. Geometrical shapes and formulas for calculating biovolume (continued)

Shape	Biovolume	Samples
	$V = \dfrac{\pi}{4} \cdot a_1 \cdot b_1 \cdot c_1 + \dfrac{\pi}{3} \cdot a_2 \cdot b_2^2$	*Climacodium* sp.
	$V \cong \dfrac{\pi}{4} \cdot a \cdot b \cdot c$	*Caloneis* sp.
	$b_2 = b_3 = b_4$ $V = \dfrac{\pi}{4} \cdot a_2 \cdot b_2^2 \cdot + \dfrac{\pi}{12} \cdot$ $\left(a_3 + a_4\right) \cdot b_2^2 + \dfrac{\pi}{6} \cdot a_1 \cdot b_1 \cdot b_2$	*Staurastrum* sp.
	$V = \dfrac{\pi}{12} \cdot a^3$	*Cosmarium* sp.

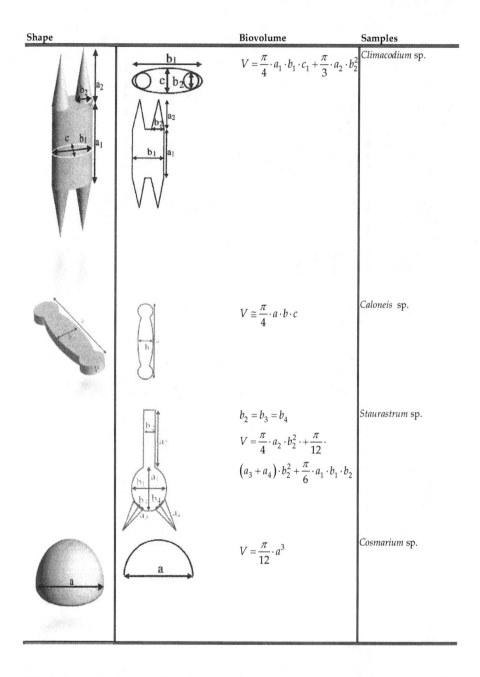

Table 1. Geometrical shapes and formulas for calculating biovolume (continued)

Shape		Biovolume	Samples
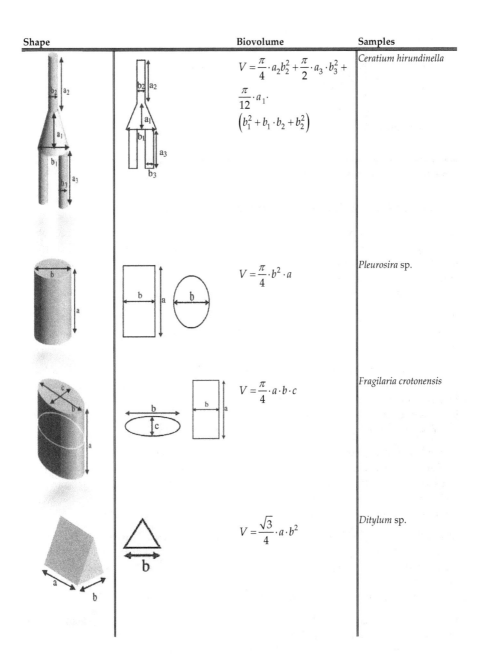		$V = \dfrac{\pi}{4} \cdot a_2 b_2^2 + \dfrac{\pi}{2} \cdot a_3 \cdot b_3^2 + \dfrac{\pi}{12} \cdot a_1 \cdot \left(b_1^2 + b_1 \cdot b_2 + b_2^2 \right)$	*Ceratium hirundinella*
		$V = \dfrac{\pi}{4} \cdot b^2 \cdot a$	*Pleurosira* sp.
		$V = \dfrac{\pi}{4} \cdot a \cdot b \cdot c$	*Fragilaria crotonensis*
		$V = \dfrac{\sqrt{3}}{4} \cdot a \cdot b^2$	*Ditylum* sp.

Table 1. Geometrical shapes and formulas for calculating biovolume (continued)

Shape	Biovolume	Samples
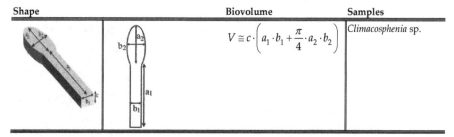	$$V \cong c \cdot \left(a_1 \cdot b_1 + \frac{\pi}{4} \cdot a_2 \cdot b_2 \right)$$	*Climacosphenia* sp.

Table 1. Geometrical shapes and formulas for calculating biovolume (continued)

7. Case studies conducted to phytoplankton biomass

There is a significant correlation between biomass of phytoplankton with the concentration of phosphorus. Changes are seen in phytoplankton biomass or production rate with the changes of the concentration of phosphorus in the lakes. It has been showed in the field studies that light and temperature play a significant role in the relationship between the extinction rate and biomass and carrying capacity of the composition of species [34].

In a study conducted in Lake Erie, samples have been collected from three different points in spring, summer and fall season for five years; 49 species were determined at the end of the study [35].

In a lake with 25000 km² surface area, phytoplankton biomass showed local changes and it was determined as 1.88 ± 0.12 g/m³, 1.04 ± 0.07 g/m³ and 0.63 ± 0.071 g/m³ in west, mid and east part of the lake respectively. It was determined that algal biomass decreased and biomasses of *Aphanizomenon flos-aque, Stephanodiscus binderounus, S. niagarae, S. tenuis, Rhodomonas minuta* decreased in the rate of 70-98% from 1970 to 1983-1987 [35].

Phytoplankton communities and distributions were investigated from the samples taken weekly from two dam lake with different nutritient levels in Sicily. Lake Arancio is a shallow eutrophic lake and Lake Rosamarina is a deep mezotrofik lake. It was stated that the increment in the concentration of nutrients in Lake Arancio doesn't change the composition of phytoplankton but increase biomass of phytoplankton [36].

In a study examining 27 lakes in Russia; plankton biomass and total phosphorus concentration were investigated and it is stated that total phosphorus concentration changes between 10-137 mg/m³ and biomass changes between 0.4-20 g/m³. Total 160 phytoplankton species were identified and it was reported that most of those are belong to the blue-green algae and euglenophyceae classes. The lakes were determined to be hypertrophic and acidic [37].

In a study conducted with Lake Dorani, it was determined that the most common class in the lake was chlorophyceae followed by cyanophyceae. Total phytoplankton biomass is found similar to eutrophic lakes. While, nanoplankton biomass constitute 90% of total phytoplankton biomass in spring but it is 10% throughout the year. It is found that total biomass is high in summer, low in winter and changes between 0.43-30.30 mg/l [38].

Seasonal changes of phytoplankton communities in Lake Managua are investigated, it is reported that blue-green algae are dominant during the research period. Seasonal biomass are measured monthly for two years and the lowest phytoplankton biomass was found at the end of the rainy seasons (October, November). In short term studies (3-14 days), important changes in biomass were reported. Nutrient levels of the lake were estimated as hypertrophic according to chlorophyll *a* value (79 µg/l yearly average, 1987-1988) [39].

During the research conducted with Lake Beysehir, diatoms and green algae are found dominant. throughout the research *Aulacoseira granulata* and *Cyclotella meneghiniana* from centric diatom, *Asterionella formosa, Cocconeis placentula, Cymbella affinis* and *Ulnaria acus* from pennate diatoms, *Monoraphidium* spp., *Mougeotia* sp. and *Scenedesmus linearis* from Chlorophyta, *Dinobryon divergens* from Chrysophyta and *Cryptomonas marssonii, Rhodomonas lacustris* from Cryptophyta, *Merismopedia glauca* from Cyanophyta are commonly found and partly numerical increases are observed. Phytoplankton biomass in the lake changes between 0.40±0.11 and 6.43±1.00 mg/l. The lake is in mesotrophic nutrients level according to average phytoplankton biomass (1.98±0.2 mg/l) and it is in good ecological quality class [40].

8. Results and suggestions

When the comparison was made between geometrical and other models, geometrical model is used more. Trials for other models in computer environment still continue. Techniques used for calculating biomass have some advantages and disadvantages. For example, *Strombomonas gibberosa* is phytoplankton with complex shape. For this reason, some different opinions arise for choosing proper geometric shape for calculating biovolume.

Three problem stands out in the estimation of biomass.

1. The shapes of phytoplankton cells has irregular and complex structure which makes it hard to measure them under microscope
2. Cell dimensions changes in the study of dead cells. In addition, it makes hard to determine chloroplast and vacuollarin.
3. Physiological state of a cell (light, temperature, nutrient) may affect cell height and intracellular volume [14].

Calculating biomass is important for determining ecological status of aquatic ecosystems. There is a relationship between cell structure of phytoplankton communities and many Physico-chemical parameters. For this reason, physical and chemical changes of water have to be considered in biomass calculaions.

Since some species show physiological changes with the changes in environmental conditions, characteristics of phytoplankton groups should be well known. In some species in group of cyanobacteria, fringes observed depending on the increase of the value of nitrogen and phosphorus in the medium. This causes changes in cell dimensions. For this reason, it is suggested to support biomass usage for classifying freshwater ecological systems with physico-chemical parameters.

Author details

Ozden Fakioglu

Atatürk University, Turkiye

Acknowledgement

The author would like to acknowledge Prof. Nilsun DEMİR and Prof. Muhammed ATAMALP for their insightful comments on this chapter.

9. References

[1] Thunmark S (1945) Zur Soziologie Des Su¨ Sswasserplankton. Eine Methodisch-O¨ Kologische Studie. Folia Limnologica Skandinvaica. 3; 1–66.

[2] Nygaard G (1949) Hidrobiological studies on some Danish ponds and lakes. Pert II: The quotient hypothesis and some little known plankton organisms. Kongelige Danske Videnskabernes Selskab Biologiske Skrifter 7; 1-293.

[3] Lepistö L., Rosenström U (1998) The most typical phytoplankton taxa in four lakes in Finland. Hydrobiologia, 369/370; 89–97.

[4] Padisak J, Grigorszky I, Borics G, Soroczki-Pinter E (2006) Use of phytoplankton assemblages for monitoring ecological status of lakes within the Water Framework Directives: the assemblage index. Hydrobiologia, 1-14.

[5] Sommer U (1986) The periodicity of phytoplankton in Lake Constance (Bodensee) in comparison to other deep lakes of central Europe. Hydrobiologia, 138; 1–7.

[6] Scheffer M, Reinaldi S, Huisman J, Weissing FJ (2003) Why plankton communities have no equilibrium: solutions to the paradox. Hydrobiologia, 491; 9- 18.

[7] Reynolds CS (1980) Phytoplankton assemblages and their periodicity in stratifying lake systems. Holarctic Ecology, 3; 141–159.

[8] Reynolds CS (1997) Vegetation processes in the pelagic. Ecology Institute, Oldendorf/Luhe, Germany.

[9] Mason CF (1991) Biology of Freshwater Pollution. 2nd ed. Longman, 351 s., Great Britain.

[10] Reynolds C, Dokulil M, Padisak J (2000) Understanding the assembly of phytoplankton in relation to the trophic spectrum: where are now? Hydrobiologia, 424; 147-152.

[11] Ediger D, Soydemir N, Kideys AE (2006) Estimation of Phytoplankton Biomass Using HPLC Pigment Analysis in the South-western Black Sea. Deep-sea Research II. 53: 1911-1922.

[12] Sun J, Liu D (2003) Geometric models for calculating cell biovolume and surface area for phytoplankton. Journal of Phytoplankton Research. 25; 1331-1346.

[13] Kayaalp GT, Polat S (2001) Tüm Gözlemli ve Eksik Gözlemli Regresyon Modelinde Klorofil-a miktarının tahmini. E.Ü. Su Ürünleri Dergisi, 18; 529-535.

[14] Sicko-Goad L, Stoermer EF, Ladewski BG (1977) A Morphometric Method for Correcting Phytoplankton Cell Volume Estimates. Protoplasma. 93: 147-163.

[15] Rowe DM (1997) Developments in Holographic-Based Scanner Desings, SPIE Conference on Optical Scanning System. San Diego, CA.

[16] Hillebrand H, Dürselen CD, Kırschtel D, Pollıngher U, Zohary T (1999) Biovolume calculation for pelagic and bentic microalgae, *J. Phycol.*, 35, 403-424.

[17] Strıckland JDH, Parssons TR (1972) A Practical Handbook Of Seawater Analysis. 2nd Ed. Bull. Fish. Res. Board. Can., p311, Canada.

[18] Wetzel RG (2001) Limnology. 3nd Edition, Saunders College, 767 p., Philadelphia.

[19] Hustedt F (1930) Bacillariophyta (Diatomeae). Heft 10. In: A. Pascher (Ed), Die Süsswasser-Flora Mitteleuropas, Verlag von Gustav Fisher, Jena, p466.

[20] Huber-Pestalozzi G (1938) Das Phytoplankton des Süsswassers, 1 Teil. Blaualgen, Bakterien, Pilze. In: A. Thienemann (Ed), Die Binnengewasser, E. Schweizerbart'sche Verlagsbuchhandlung, 342 p., Stuttgart.

[21] Huber-Pestalozzi G (1942) Das Phytoplankton des Süsswassers, 2 Teil. Diatomeen. In: A. Thıenemann (Ed), Die Binnengewasser, E. Schweizerbart'sche Verlagsbuchhandlung, p549., Stuttgart.

[22] Huber-Pestalozzi G (1950) Das Phytoplankton des Süsswassers, 3 Teil. Cryoptophyceen, Chloromonadien, Peridineen. In: A. Thienemann (Ed), Die Binnengewasser, E. Schweizerbart'sche Verlagsbuchhandlung, 310 p., Stuttgart.

[23] Cox EJ (1996) Identification of Freshwater Diatoms from Live Material. Chapman and Hall, p158, London.

[24] Prescott GW (1973) Algae of the Western Great Lakes Area, 5th ed. WM. C. Brown Co. Publ, 977, Dubuque.

[25] Lind ME, Brook AJ (1980) A key to the Commoner Desmids of the English Lake District. Freswater Biol. Assoc. Publ., 123, Cumbria.

[26] Komarek J, Fott B (1983) Chlorococcales, 7. Teil. 1Halfte. In: J. Elster and W. Ohle (Eds), Das Phytoplankton des Süsswassers, E. Schweizerbart'sche Verlagsbuchhandlung, p1043., Stuttgart.

[27] Komarek J, Anagnostidis K (1999) Cyanoprokaryota 1. Teil: Chroococcales. In: H. Ettl, G. Gartner, H. Heynig, D. Mollenhauer (Eds), Süsswasserflora von Mitteleuropa, Spektrum Akademischer Verlag, p548., Heidelberg.

[28] John DM, Whitton BA, Brook AJ (2002) The Freshwater Algal Flora of The British Isles. Cambridge Univ. Press, Cambridge, p702.

[29] Popovski J, Pfiester LA (1990) Dinophyceae (Dinoflagellida), Band 6. In: H. Ettl, J. Gerloff, H. Heynig, D. Mollenhauer (Eds). Süsswasserflora von Mitteleuropa, Gustav Fishre Verlag, 243, Jena.

[30] Utermöhl H (1958) Zur Vervolkommnung deer quantitativen Phytoplankton-Methodik. Mitteilungen der Internationale Vereinigung der theoretretische und Angewandte Limnologie, 5; 567-596.

[31] Anonymous (1995) Standard Methods for the Examination of Water and Wastewater. 19th ed., American Public Health Association (APHA), p1193., Washington.

[32] Wetzel RG, Likens GE (1991) Limnological Analysis. 2nd ed. Springer Verlag, New York.

[33] Rott E (1981) Some results from phytoplankton counting intercalibrations. Schweiz. Z. Hydrol. 43; 34-59.

[34] Heyman U, Lundgren A (1988) Phytoplankton biomass and production in relation to phosphorus. Hydrobiologia, 170; 211-227.

[35] Makarewicz JC (1993) Phytoplankton Biomass and Species Composition in Lake Erie, 1970 to 1987. J. Great Lakes Res, 19(2); 258–274.

[36] Naselli-Flores L, Barone R (1998) Phytoplankton dynamics in two reservoirs with different trophic state (Lake Rosamarina and Lake Arancio, Sicily, Italy) Hydrobiologia, 369/370; 163-178.

[37] Trifonova IS (1998) Phytoplankton composition and biomass structure in relation to trophic gradient in some temperate and subarctic lakes of Northwestern Russia and the Prebaltic. Hydrobiologia, 369/370; 99–108,

[38] Temponeras M, Kristiansen J, Moustaka-Gouni M (2000) Seasonal variation in phytoplankton composition and physical-chemical features of the shallow Lake Doïrani, Macedonia, Greece Hydrobiologia, 424; 109–122.

[39] Hooker E, Hernandez S (2006) Phytoplankton biomass in Lake Xolotlán (Mamagua): Its seasonal and horizontal distribution. Aquatic Ecology, 25 (2); 125–131.

[40] Fakıoglu O, Demır N (2011) The Spatial and Seasonal Variations of Phytoplankton Biomass in Beyşehir Lake. Ekoloji. 80; 23-32.

Permissions

The contributors of this book come from diverse backgrounds, making this book a truly international effort. This book will bring forth new frontiers with its revolutionizing research information and detailed analysis of the nascent developments around the world.

We would like to thank Miodrag Darko Matovic, for lending his expertise to make the book truly unique. He has played a crucial role in the development of this book. Without his invaluable contribution this book wouldn't have been possible. He has made vital efforts to compile up to date information on the varied aspects of this subject to make this book a valuable addition to the collection of many professionals and students.

This book was conceptualized with the vision of imparting up-to-date information and advanced data in this field. To ensure the same, a matchless editorial board was set up. Every individual on the board went through rigorous rounds of assessment to prove their worth. After which they invested a large part of their time researching and compiling the most relevant data for our readers. Conferences and sessions were held from time to time between the editorial board and the contributing authors to present the data in the most comprehensible form. The editorial team has worked tirelessly to provide valuable and valid information to help people across the globe.

Every chapter published in this book has been scrutinized by our experts. Their significance has been extensively debated. The topics covered herein carry significant findings which will fuel the growth of the discipline. They may even be implemented as practical applications or may be referred to as a beginning point for another development. Chapters in this book were first published by InTech; hereby published with permission under the Creative Commons Attribution License or equivalent.

The editorial board has been involved in producing this book since its inception. They have spent rigorous hours researching and exploring the diverse topics which have resulted in the successful publishing of this book. They have passed on their knowledge of decades through this book. To expedite this challenging task, the publisher supported the team at every step. A small team of assistant editors was also appointed to further simplify the editing procedure and attain best results for the readers.

Our editorial team has been hand-picked from every corner of the world. Their multi-ethnicity adds dynamic inputs to the discussions which result in innovative outcomes. These outcomes are then further discussed with the researchers and contributors who give their valuable feedback and opinion regarding the same. The feedback is then collaborated with the researches and they are edited in a comprehensive manner to aid the understanding of the subject.

Apart from the editorial board, the designing team has also invested a significant amount of their time in understanding the subject and creating the most relevant covers. They scrutinized every image to scout for the most suitable representation of the subject and create an appropriate cover for the book.

The publishing team has been involved in this book since its early stages. They were actively engaged in every process, be it collecting the data, connecting with the contributors or procuring relevant information. The team has been an ardent support to the editorial, designing and production team. Their endless efforts to recruit the best for this project, has resulted in the accomplishment of this book. They are a veteran in the field of academics and their pool of knowledge is as vast as their experience in printing. Their expertise and guidance has proved useful at every step. Their uncompromising quality standards have made this book an exceptional effort. Their encouragement from time to time has been an inspiration for everyone.

The publisher and the editorial board hope that this book will prove to be a valuable piece of knowledge for researchers, students, practitioners and scholars across the globe.

List of Contributors

Edmilson José Ambrosano, Raquel Castellucci Caruso Sachs and Juliana Rolim Salomé Teramoto
APTA, Pólo Regional Centro Sul, Piracicaba (SP), Brazil

Fábio Luis Ferreira Dias
APTA, Pólo Regional Centro Sul, Piracicaba (SP), Brazil
Instituto Agronômico (IAC), Campinas (SP), Brazil

Heitor Cantarella
Instituto Agronômico (IAC), Campinas (SP), Brazil

Gláucia Maria Bovi Ambrosano
Universidade Estadual de Campinas, Departamento de Odontologia SocialPiracicaba (SP) Brazil

Eliana Aparecida Schammas
Instituto de Zootecnia, Nova Odessa (SP), Brazil

Fabrício Rossi
Faculdade de Zootecnia e Engenharia de Alimentos (FZEA/USP), Pirassununga, (SP), Brazil

Paulo Cesar Ocheuze Trivelin and Takashi Muraoka
Centro de Energia Nuclear na Agricultura (CENA/USP), Piracicaba (SP), Brazil

Rozario Azcón
Estação Experimental de Zaidin, Granada, Espain

Qingwu Xue
Texas A&M AgriLife Research and Extension Center at Amarillo, Amarillo, TX, USA

Guojie Wang and Paul E. Nyren
North Dakota State University, Central Grasslands Research Extension Center, Streeter, ND, USA

Miled El Hajji
ISSATSO, Université de Sousse, cité taffala, 4003 Sousse, Tunisie and LAMSIN, BP. 37, 1002 Tunis-Belvédère, Tunis, Tunisie

Alain Rapaport
UMR INRA-SupAgro 'MISTEA' and EPI INRA-INRIA 'MODEMIC' 2, place Viala, 34060 Montpellier, France

Jian Yu, Michael Porter and Matt Jaremko
Hawaii Natural Energy Institute, University of Hawaii at Manoa, Honolulu, Hawaii, USA

Viktor J. Bruckman and Gerhard Glatzel
Austrian Academy of Sciences (ÖAW), Commission for Interdisciplinary Ecological Studies (KIOES), Vienna, Austria

Shuai Yan
Northwest Agricultural and Forestry University, College of Forestry, Yangling, China

Eduard Hochbichler
University of Natural Resources and Life Sciences (BOKU), Institute of Silviculture, Vienna, Austria

Jude Liu
Department of Agricultural and Biological Engineering, The Pennsylvania State University, University Park, Pennsylvania, USA

Robert Grisso and John Cundiff
Department of Biological Systems Engineering, Virginia Tech, Blacksburg, Virginia, US

Maurício Emerenciano
Posgrado en Ciencias del Mar y Limnología, Universidad Nacional Autónoma de México (UNAM), Unidad Multidisciplinaria de Docencia e Investigación (UMDI), UNAM, Sisal, Yucatán, Mexico
Santa Catarina State University, Centro de Educação Superior da Região Sul (CERES), Laguna, Santa Catarina, Brazil

Gabriela Gaxiola
UMDI, Facultad de Ciencias, UNAM, Sisal, Yucatán, México

Gerard Cuzon
Ifremer (Institut Français de Recherche pour l'Exploitation de la Mer) Taravao, Tahiti, French Polynesia

Ozden Fakioglu
Atatürk University, Turkiye

Printed in the USA
CPSIA information can be obtained
at www.ICGtesting.com
JSHW011402221024
72173JS00003B/394

9 781632 391827